Mental Math

Tricks To Become A Human Calculator

By Abhishek V.R

Ofpad – The School of Genius

Copyright

Dedication

All that I am & all that I will ever be, I owe it to my mother.

This book is also dedicated to all members of Ofpad.com, who push the boundaries of their intelligence every day.

Contents

Foreword

I want to begin by telling you a story. This is a story that I am pretty embarrassed to share. Yet it is one you need to hear because it will reveal how you too can do math in your head.

You see I wasn't always good with numbers, and I certainly wasn't smart growing up. In fact far from it - I HATED math.

One beautiful morning, I was in the backseat of my cousin's motorcycle enjoying the cool breeze hitting my face.

My summer was coming to an end, and my cousin was dropping me at the railway station. I was returning home after enjoying the monsoon rains. Life was good until it decided to punch in the face.

This was a critical day of my life. In another 30 minutes, the results of my exam were going to be out. This was important because how much I scored in this exam determined whether or not I went to college.

In school, I wasn't very serious about my grades. My entire time was either spent playing video games or talking to my high school girlfriend on the phone.

I had no real goal or purpose. If you had asked me what I wanted to become, I would have told you, "I am still figuring that out". I was just an average guy going with the flow.

I was happy with just getting by and thought I would just make my way into college without really trying. However, life was about to give me a wakeup call.

My cousin and I reached the station, and we were waiting for the train. He was one of the few people who had a smartphone back then. While we waited for the train, my cousin offered to check how much I scored on his iPhone. He punched in my details and waited for the result to load. Seconds seemed like ages. His excited face was replaced with a look of sympathy.

The look on his face only told me that I didn't even get the bare minimum required to get into college. I grabbed his phone and saw how much I scored. My face turned pale.

I got the worst scores possible in math. It would take a miracle for me to go to college.

I later learnt that all my friends did well. My girlfriend did so well that she was going to study medicine. She broke up with me soon after. In my own eyes, everyone around me seemed a lot smarter than me.

I started loathing myself. I lost all self-esteem. Needless to say this was one of the lowest points of my life. I had hit rock bottom, and I didn't know if I even had a future. I would spend the next few months in depression crying about everything that went wrong in my life.

My parents were anxious. However, word got around, and one of my father's friends offered to give me a recommendation. Life gave me another chance, and I somehow went to college.

But this wasn't enough to restore my lost self-esteem. Everyone I knew, got better grades than me. To my young mind, grades were a measure of intelligence. By that measure, I saw myself as having one of the lowest IQs in the world.

The late president of my country once said, "Dreams are not what you have when you sleep. The true dreams are the ones that don't let you sleep."

Becoming the smartest possible version of myself became an obsession that wouldn't let me sleep. I had to become incredibly intelligent, to restore my lost self-esteem. This made me want to learn everything I possibly could that would push the boundaries of my intelligence.

I decided to start this journey from a place I hated the most – MATH.

It made a lot of sense to me back then to start here because my lack of love for it almost cost me my future. So the first thing I did was learn how to do math in my head.

I read every possible book there was. Tracked down every article and video published on the subject and absorbed it. I combined what I learnt from different places to create my unique system that would let anybody do mental math lightning fast.

The techniques can be learnt in a few minutes. When I first discovered it, it started to change the way I thought.

Since calculations were done in the head, I acquired better memory habits. My concentration and my ability to think improved. But most importantly, learning how to do mental math pushed the boundaries of what I thought was possible.

If I could now do math faster in my head after having hated it for a good part of my life, then I could do anything.

I have been relentlessly pushing the boundaries of my intelligence in every way possible ever since. Now I have an MS degree and work in the field of data analytics.

Math is what I enjoy doing the most. Nobody believes me when I tell them I was an average student in school. They think I was gifted from birth.

Remember, my brain is no different from yours when it comes to numbers. I am just a guy who happens to have a system which makes it easier to do mental math.

This is a system that allows people like us to experience a life where math and numbers become useful, as we experience a new and powerful way to think.

This is the same system that has allowed me to effortlessly calculate numbers in my head faster than a calculator after having hated math for a good part of my life.

And today, in this book, I am going to share the secret of doing mental math with you. The secret that will change your life the way it did mine.

Chapter 1 - What Is Ofpad Mental Math System?

The Ofpad Mental Math system is the only system to do math in your head faster than a calculator making math your new superpower. These radically simple techniques for mental math will work for you even if you hate math and if you are terrible at math, to begin with.

This mental math book was written after analysing the techniques used by over 27 mental math geniuses across the world, from countries including Sweden, India and China. I uncovered these hidden patterns and unique (sometimes 'odd') tactics they use to do math with superhuman speed. And now, starting today, their success can be your success.

What is remarkable about this system is that it is so simple that even a child can do it. The strategies of these math geniuses can now be duplicated easily by you, giving you the power of using math and numbers in your life. You will enjoy an improved memory and develop a laser-sharp concentration, all while you calculate faster than you ever dreamed possible.

You will not find complicated rules that work for specific situations in this book. Instead, you will learn broad concepts that you can apply to all types of math problems.

You will not find math techniques that will require you to use a paper. After you finish this book, you will start calculating in your head faster than before.

However, for the techniques to become second nature to you, you will have to do the practice exercises that come with the book.

Now, before we go any further, I need to be 100% honest with you. If what you want is a "magic pill" that you can swallow to become a genius (something you already know will never work in practice), then you can stop reading this book right now.

The Ofpad Mental Math System isn't that kind of fantasy land nonsense. It's the real deal, and it is only for folks like you who are serious about improving their intelligence, and who are willing to put in the time to practice and hone this new skill.

While doing math in your head isn't "easy" (nothing worthwhile is). With this book, I have made mental math as easy as it can get.

So, if you're like most folks reading this book, and if you're 100% ready to learn the real secrets of doing mental math faster than a calculator, while increasing concentration, developing better memory habits and learning new ways to think and do math entirely in your head faster than a calculator, then the Ofpad Mental Math System isn't just "A" solution for you. It's the "ONLY" solution for you!

You may think you have seen it all when it comes to math. But trust me you would have never seen anything like this before. This system is entirely different, and it's nothing like what you have been taught in school. The method I am about to share with you will shock you, and you will be left wondering why nobody ever taught you this when you were young.

Chapter 2 - DS Method of Checking

In this chapter, we will cover the DS Method to check your calculation. DS stands for Digit Sum.

This method has been known to mathematicians for several centuries. But it is not widely known or taught in school. So it is not used much in everyday life.

You will be able to check the answers of your addition, subtraction, multiplication and division problems quickly using this DS Method. We are covering this first so that you can use it throughout this book.

Rules To Calculate Digit Sum

Rule 1 - Digit sum is simply the sum of all digits in a number.

For example, the digit sum of 213 is 6. You get this by adding 2 + 1 + 3 which gives 6.

Number: 213
Digit Sum: 2 + 1 + 3 = 6

Rule 2 - Digit sum should be a single digit number.

For example, the digit sum of 2134 is 1. You get this by adding 2 + 1 + 3 + 4 which gives 10. Since a digit sum is a single number we also add 1 + 0 to give 1, and 1 becomes the digit sum of 2134.

Number: 2134
Digit Sum: 2 + 1 + 3 + 4 = 10
Single Digit: 1 + 0 = 1

Rule 3 - You should ignore the number nine when adding across.

Number: 909
Digit Sum: 0 (Ignoring 9s we get 0)

Rule 4 - You can also ignore the digits which add up to 9 (e.g. 1 & 8 or 3, 4 & 2). Your digit sum will be the same irrespective of whether or not you ignore the individual numbers.

Dropping 1 & 8 which adds up to 9 we get the digit sum as 2.

Number: 1802
Digit Sum: 0 + 2 = 2

Alternatively, without dropping 1 & 8 which adds up to 9, we will still get the same digit sum 2.

Number: 1802
Digit Sum: 1 + 8 + 0 + 2 = 11
Single Digit: 1 + 1 = 2

So your digit sum remains the same irrespective of whether or not you drop the numbers that adds up to 9.

It is okay to drop 9 because digit sum will remain unchanged when you add any number to 9. If you see the additions below, the number you add to 9 is the digit sum of the answer.

$$1 + 9 = 10 = 1 + 0 = 1$$
$$2 + 9 = 11 = 1 + 1 = 2$$
$$3 + 9 = 12 = 1 + 2 = 3$$
$$4 + 9 = 13 = 1 + 3 = 4$$
$$5 + 9 = 14 = 1 + 4 = 5$$
$$6 + 9 = 15 = 1 + 5 = 6$$
$$7 + 9 = 16 = 1 + 6 = 7$$
$$8 + 9 = 17 = 1 + 7 = 8$$
$$9 + 9 = 18 = 1 + 8 = 9$$

Look at the numbers in **bold** in the far left and the far right. They are the same. So by dropping 9 or numbers that add up to 9, you will skip the intermediate addition that you will end up doing in between the two bold numbers. You will skip an unnecessary step during your calculation of the digit sum.

Rule 5 - Decimals work the same way as normal numbers. The decimal point will not affect the digit sum.

Practice

Let us practice finding the digit sum with a few examples.

First, we learnt that digit sum is a sum of all digits in a number.

Find the digit sum of 122.

Number: 122
Digit Sum: 1 + 2 + 2 = 5

Second, we also learnt, the digit sum should be a single digit number.

Find the digit sum of 671.

Number: 671
Digit Sum: 6 + 7 + 1 = 14

Since digit sum should be a single number, we add 1 and 4 to get 5.

Single Digit: 1 + 4 = 5

We learnt that in adding across a number, you should ignore all 9s.

Find the digit sum of 9999.

Number: 9999
Digit Sum: 0

Since all the numbers are 9, we will ignore every number to get 0.

We learnt that you could ignore the numbers which add up to 9, but you will get the same digit sum even if you don't ignore them.

Find the digit sum of 1723.

Number: 1723
Digit Sum: 1 + 3 = 4 (Ignoring 7 + 2)

We should add 1 + 7 + 2 + 3. But we can ignore 7 + 2 since it adds up to 9. So the digit sum becomes 1 + 3 = 4.

Alternatively: 1 + 7 + 2 + 3 = 13
Single Digit: 1 + 3 = 4

If instead we added all the numbers 1 + 7 + 2 + 3 we would get 13. Since digit sum should be a single digit number, we add 1 + 3 to get 4. This is the same digit sum we would have got had we ignored 7 and 2 which adds up to 9.

Rules To Apply DS Method of Checking

The rule for checking your answers with the DS Method is simple:

1. Whatever you do to the numbers (add, subtract, multiply or divide), also do to the digit sum of the numbers.
2. The result you get from the digit sum of the numbers should be equal to the digit sum of the answer.

Let us look at an example to understand the method.

Checking Addition

Let us check if the sum of 93 + 11 = 104 without explicitly calculating it.

Whatever you do to the numbers, also do to the digit sum of the numbers.

Problem: 93 + 11 = 104
Digit Sum: 3 + 2 = 5

Dropping 9 from 93 we get the digit sum of 93 as 3.

Adding the two digits of 11, we get the digit sum of 11 as 2.

Adding 3 + 2 we get 5.

The result of the addition is correct because 5 is also the digit sum of the answer 104 (got by adding 1 + 0 + 4 = 5).

Checking Subtraction

Let us check if the difference between 93 - 11 = 82 without explicitly calculating it.

Whatever you do to the numbers, also do to the digit sum of the numbers.

Problem: 93 - 11 = 82
Digit Sum: 3 - 2 = 1

So dropping 9 from 93 we get the digit sum of 93 as 3.

Adding the two digits of 11, we get the digit sum of 11 as 2.

Subtracting 3 - 2 we get 1.

The result of the subtraction is correct because the digit sum of the answer 82 is also 1 (got by adding 8 + 2 = 10 = 1 + 0 = 1).

When Digit Sum is Negative for Subtraction

Sometimes the subtraction of the digit sum results in a negative number.

If the subtraction of the digit sums is negative add 9 to the negative number.

Let us look at an example.

Let us check if the difference between 23 - 17 is equal to 6 without explicitly calculating it.

Whatever you do to the numbers, also do to the digit sum of the numbers.

Problem: 23 - 17 = 6
Digit Sum: 5 - 8 = -3 (+9) = 6

So adding 2 and 3 in 23 we get the digit sum of 23 as 5.

Adding the two digits of 17, we get the digit sum of 17 as 8.

Subtracting 5 - 8 we get -3.

If the subtraction of the digit sums is negative, add 9 to the negative number.

So adding 9 to -3 we get 6.

The digit sum of the answer is also 6, so the subtraction is done correctly.

Checking Multiplication

Let us check if 93 multiplied by 11 is equal to 1023 without explicitly calculating it.

Whatever you do to the numbers, also do to the digit sum of the numbers.

Problem: 93 x 11 = 1023
Digit Sum: 3 x 2 = 6

So dropping 9 from 93 we get the digit sum of 93 as 3.

Adding the two digits of 11, we get the digit sum of 11 as 2.

Multiplying 3 by 2 we get 6.

The result of the multiplication is correct because 6 is also the digit sum of the answer 1023 (got by adding $1 + 0 + 2 + 3 = 6$).

Checking Division

Let us check if the division of 110 by 11 is equal to 10 without explicitly calculating it.

Whatever you do to the numbers, also do to the digit sum of the numbers.

Problem: 110 / 11 = 10
Digit Sum: 2 / 2 = 1

Adding the three digits of 110 we get the digit sum of 110 as 2.

Adding the two digits of 11, we get the digit sum of 11 as 2.

Dividing 2 by 2 we get 1.

The digit sum of the answer 10 is also 1 (got by adding 1 + 0 = 1), so the division is done correctly.

Tip – Reduce Running Total to Single Digits

Reduce the digit sum to a single figure as you go along and don't wait until the end to reduce it to a single digit number.

Suppose you are finding the digit-sum of 512,422.

Start on the left and add across: 5 + 1 + 2 and so on. Say to yourself only the running totals.

Totals: 6, 8, 12 and reduce to a single figure as you go along. Reduce this 12 to 3 (got by adding 1 + 2).

Go ahead with this 3 and add the remaining two digits of our example:

3, 5, 7. The digit-sum is 7.

This is less difficult than reducing the digit sum to a single digit towards the end as follows:

5 + 1 + 2 + 4 + 2 + 2 = 16, then 1 + 6 equals 7.

In a very long number, reducing the running total to a single digit number saves you a lot of time.

Summary

This DS method of checking will be useful in the coming chapters. It can also be used in any practice problems that you do. Practice will make using these techniques effortless and easy. Students of the video course **ofpad.com/mathcourse** can post their questions in the discussion section or you can email me at **hq@ofpad.com**.

Chapter 3 – Mental Math Video Course

Before I tell you about the video course, I want to assure you that after you finish reading this book, you will have everything you need to be a mental math genius. That said, I know for a fact that someone eager to improve their ability to do mental math will want to take it a step further. The mental math tricks in this book were first available in the Ofpad Mental Math Course which is delivered in video format. The course is open for students to enrol in this link: **ofpad.com/mathcourse**

The video course has already helped over 10,000 students from over 132 countries. This book was written to help reach even more students who consume information in the traditional book format.

You want to calculate fast like a math genius, so you would also want to learn the right way to visualise numbers in your head. In the video course, the techniques & the mental steps are explained with the help of animations. When the students of the video course do the math later, they visualise the mental process the same way they see in the video.

When writing this book, I felt handicapped because I could now only use static text and not animations. Even though the techniques and examples are the same in both the book and the course, there will still be a difference between the students who just read the book and the students who have taken the video course. This is because the way the video course students visualise the techniques in their mind will be very different.

Students of the video course also benefit from being able to ask questions if they don't understand something. When a student needs some clarification, then one of the other students in the community or I will be able to immediately answer all queries in the discussion section of the video course.

If you are an auditory or visual learner who retains information by hearing and visualising things, then you will benefit from enrolling in the video course. However, if you are someone who prefers consuming information in text format, then this book might be all that you need. But, you might have to read and re-read the same sections multiple times to completely absorb and retain everything. This is entirely normal, and I recommend you read this book as many times as you think is necessary to understand everything completely.

You might take a few weeks to finish this book. Readers who enrol in the video course, however, will finish everything in a couple of hours and can use this book to reinforce what they learn.

If you enrol in the video course now, you will be one of the founding students. That means a much lower price for you today because frankly, I want you to benefit from taking the course and I refuse to let finances stop you.

You have already purchased this book at the discounted price. Also because you bought this book, it means you will not be paying the retail value of $297 for the video course today.

You are not even going to be paying half the price of $197.

Your total investment today is now only one payment of just $45.

This price is available only for our founding students, and when we have enough students, the price will go up.

Visit **ofpad.com/mathcourse** and purchase the course so you lock in your discounted price today of only $45 or you risk coming back at a later date or even a few minutes from now to see the price go up.

I know this might still be a significant investment for you and it can be a bummer to buy something that does not work. Believe me when I say this, during my personal growth journey I have been there many times.

So you don't have to decide right now. Take advantage of the unconditional "Triple Guarantee". Just try the video course for 30 days and:

1) If you are not calculating with the speed you expect at the end of 30 days or
2) If you do not start calculating faster in 48 hours after completing the video course or
3) If you are not happy with the video course for any other reason

Then just let me know and I will say thank you for trying and will refund every penny you invest in purchasing the video program. No questions asked, no hassles, and no hard feelings.

I am sure that is not going to happen. But I just want you to know that the option is there, so your investment today is going to be 100% risk-free.

Remember that it's not just a fantastic bargain you'll be receiving today. You'll be joining in on one of the biggest movements in history and the ever-growing Ofpad community.

Not to mention the fact you'll make a whole new group of friends that will help you progress, stay inspired, and have more fun along the road to success!

Remember, unlike other things you have seen in the past, this is the real deal, and it's 100% guaranteed. Try the Ofpad Mental Math Video Course for 30 full days and see the results you desire, or you pay nothing!

Those who act now will lock in their discount and this free bonus. Others won't be so fortunate. Visit **ofpad.com/mathcourse** to claim your copy of the video course with the incredible discount & 30-day 100% money back guarantee.

Take advantage of this limited time offer, while you still can. Visit **ofpad.com/mathcourse** & claim your copy of the video course and get on the fast track to becoming a human calculator using the smartest, proven system for doing math in your head.

Chapter 4 - Introduction to Mental Math

This is an introductory chapter that will tell you how the rest of the book is structured and will also give you a taste of what it is like to do mental maths.

We will cover generic techniques to add, subtract, multiply and divide any set of numbers in the next set of chapters. However, in this chapter, you will first learn a few mental habits you need to do mental math. We will do this by looking at the technique to multiply by 11. The general techniques to multiply will be covered in the coming chapter.

Course Structure

Each chapter in this book will have the following structure:

Section 1 - First, the steps of the mental math technique will be explained.

Section 2 - Then we will apply the fast math technique to 3 different examples to illustrate each step of the method discussed earlier. Each example will have some variation so you will be able to understand how the method is applied completely.

Section 3 - After the three examples which illustrates the mental math method, you will be given practice exercises so that you can apply these techniques yourself and master them. At the end of the book, I will tell you how you can get practice workbooks every week, so you will be able to practice until these techniques become second nature to you.

Throughout this book, there might come a point in time where you might have questions, queries, clarifications, suggestions or feedback. Students of the Ofpad Mental Math Video Course

ofpad.com/mathcourse can post this in the discussion section, and it will be answered by one of the other students or me.

Defining a Multiplicand & a Multiplier

Before we get into mental maths techniques, let us quickly define what a multiplicand and a multiplier are.

Take the following example:

43 x 23

43 is the multiplicand. It is the number being multiplied.

23 is the multiplier. It is the number which is multiplying the first number.

The description of the steps in the mental math technique will refer to the numbers in a multiplication as multiplicand and multiplier. So if you are clear on which number is the multiplicand and which number is the multiplier, you will be able to understand the description better. Don't worry if you don't remember it though because the examples illustrating the technique will clear any confusion you might have.

Multiplication By 11

As an introduction to mental math, let us now look at the technique to multiply by 11, along with its three examples illustrating the method.

There are three steps to multiply by 11:

Step 1 - The first number of the multiplicand (number multiplied) is put down as the left-hand number of the answer.

Step 2 - Each successive number of the multiplicand is added to its neighbour to the left.

Step 3 - The last number of the multiplicand becomes the right-hand number of the answer.

Let us look at an example illustrating this method.

Example 1
Let us multiply.

$$\begin{array}{r} 423 \\ \underline{\times\ 11} \end{array}$$

Step 1 - The first number of the multiplicand (number multiplied) is put down as the left-hand number of the answer.

$$\begin{array}{r} 423 \\ \underline{\times\ 11} \\ 4\ \_\ \_\ \_ \end{array}$$

So we put down 4 as the left-hand digit of the answer.

Step 2 - Each successive number of the multiplicand is added to its neighbour to the left.

So 4+2 gives 6, and 6 becomes the second digit of the answer.

$$\begin{array}{r} 423 \\ \underline{\times\ 11} \\ 4\ 6\ \_\ \_ \end{array}$$

Moving on to the next two digits, we add 2 + 3 which gives 5, and 5 becomes the next digit of the answer.

$$423$$
$$\underline{\times\ 11}$$
$$4\ 6\ 5\ \_$$

Step 3 - The last number of the multiplicand becomes the right-hand number of the answer.

So we put down 3 as the last digit of the answer.

$$423$$
$$\underline{\times\ 11}$$
$$4\ 6\ 5\ 3$$

The answer is 4653.

Example 2

Let us try another example. Multiply the following:

$$534$$
$$\underline{\times\ 11}$$
$$\_\ \_\ \_\ \_$$

Step 1 - The first number of the multiplicand is put down as the left-hand number of the answer.

So we put down 5 as the left-hand digit of the answer.

$$534$$
$$\underline{\times\ 11}$$
$$5\ \_\ \_\ \_$$

Step 2 - Each successive number of the multiplicand is added to its neighbour to the left.

So 5+3 gives 8, and 8 becomes the second digit of the answer.

534
x 11
5 8 _ _

Moving to the next two digits, we add 3 + 4 which gives 7, and 7 becomes the third digit of the answer.

534
x 11
5 8 7 _

Step 3 - The last number of the multiplicand becomes the right-hand number of the answer.

So we put down 4 as the last digit of the answer.

534
x 11
5 8 7 4

The answer is 5874.

Example 3
Let us try another example. Multiply 726 by 11.

726
x 11

_ _ _ _

Take a second to apply the technique by yourself as fast as you can. Once you have the answer, you can check the steps below to see if you got your answer right.

Step 1 - The first number of the multiplicand is put down as the left-hand number of the answer.

So we put down 7 as the left-hand digit of the answer.

$$726$$
$$\underline{\times 11}$$
$$7\,\_\,\_\,\_$$

Step 2 - Each successive number of the multiplicand is added to its neighbour to the left.

So 7+2 gives 9, and 9 becomes the second digit of the answer.

$$726$$
$$\underline{\times 11}$$
$$7\,9\,\_\,\_$$

Moving to the next two digits, we add 2 + 6 which gives 8, and 8 becomes the third digit of the answer.

$$726$$
$$\underline{\times 11}$$
$$7\,9\,8\,\_$$

Step 3 - The last number of the multiplicand becomes the right-hand number of the answer.

So we put down 6 as the last digit of the answer, and the final answer is 7986.

$$726$$
$$\underline{\times 11}$$
$$7\,9\,8\,6$$

If you got your answer wrong, don't worry. Just revisit the techniques and examples we covered in this chapter.

Carrying Over In Mental Math
Now when you add the numbers, and if it results in a sum which has two digits, you will have to carry over the first digit.

So here is the complete rule for multiplication by 11 incorporating the step of carrying over.

Step 1 - The first number of the multiplicand is put down as the left-hand number of the answer.

Step 2 - Each successive number of the multiplicand is added to its neighbour to the left.

Step 3 - If the addition results in two figures, carry over the 1 (Note: The two figure number will never be more than 18 (got by adding 9+9)).

Step 4 - The last number of the multiplicand becomes the right-hand number of the answer.

Note: Only the third step is new. The rest remains the same.

Example 1
Let us look at an example. Multiply 619 by 11.

$$619$$
$$\underline{\times 11}$$

$$----$$

Step 1 - The first number of the multiplicand is put down as the left-hand number of the answer.

So we put down 6 as the left-hand digit of the answer.

$$619$$
$$\underline{\times 11}$$
$$6\ \_\ \_\ \_$$

Step 2 - Each successive number of the multiplicand is added to its neighbour to the left.

So 6+1 gives 7, and 7 becomes the second digit of the answer.

$$\begin{array}{r} 619 \\ \underline{\times\ 11} \\ 6\ 7\ \_\ \_ \end{array}$$

Moving to the next two digits, we add 1 + 9 which gives 10.

$$\begin{array}{r} 619 \\ \underline{\times\ 11} \\ 6\ 7\ \_\ \_ \\ 1\ 0 \end{array}$$

Step 3 - If the addition results in two figures, carry over the 1.

The second digit of the sum, which is 0 becomes the third digit of the answer.

$$\begin{array}{r} 619 \\ \underline{\times\ 11} \\ 6\ 7\ 0\ \_ \\ 1\ \uparrow \end{array}$$

Since the addition resulted in two digits, we carry over the one, so the 7 now becomes 8.

$$\begin{array}{r} 619 \\ \underline{\times\ 11} \\ 6\ 8\ 0\ \_ \end{array}$$

Step 4 - The last number of the multiplicand becomes the right-hand number of the answer.

So we put down 9 as the last digit of the answer. The final answer is 6809.

619
x 11
6 8 0 9

This concept of carrying over applies to all mental math techniques that we will see in the future chapters. We are looking at this as a separate step in this first chapter for understanding and clarity. However, in the later chapters, we will skip describing this as a separate step.

Example 2

Let us look at another example. Multiply 348 x 11.

348
x 11

_ _ _ _

Step 1 - The first number of the multiplicand is put down as the left-hand number of the answer.

So we put down 3 as the left-hand digit of the answer.

348
x 11
3 _ _ _

Step 2 - Each successive number of the multiplicand is added to its neighbour to the left.

So 3+4 gives 7. This becomes the second digit of the answer.

348
x 11
3 7 _ _

Moving to the next two digits, we add 4 + 8 which gives 12.

$$\begin{array}{r} 348 \\ \underline{\times\ 11} \\ 3\ 7\ \_\ \_ \\ 1\ 2 \end{array}$$

Step 3 - If the addition results in two figures, carry over the 1.

The second digit of the sum, which is 2 becomes the third digit of the answer.

$$\begin{array}{r} 348 \\ \underline{\times\ 11} \\ 3\ 7\ 2\ \_ \\ 1\ \uparrow \end{array}$$

Since the addition resulted in two digits, we carry over the one, so the 7 now becomes 8.

$$\begin{array}{r} 348 \\ \underline{\times\ 11} \\ 3\ 8\ 2\ \_ \end{array}$$

Step 4 - The last number of the multiplicand becomes the right-hand number of the answer.

So we put down 8 as the last digit of the answer.

$$\begin{array}{r} 348 \\ \underline{\times\ 11} \\ 3\ 8\ 2\ 8 \end{array}$$

The answer is 3828.

If carrying numbers over seems to strain your mental faculties a bit, do not worry. Carrying over numbers will become effortless and easy the more your practice and progress through this book. This might probably be the first time you are trying to carry over the numbers in your head, so it will take a bit of getting used to.

Example 3

Let us look at another example. Multiply 428 by 11.

$$\begin{array}{r} 428 \\ \underline{\times 11} \\ \_\ \_\ \_\ \_ \end{array}$$

Take a second to apply the technique by yourself as fast as you can. Once you have the answer, you can check the steps below to see if you got your answer right.

Step 1 - The first number of the multiplicand (number multiplied) is put down as the left-hand number of the answer.

So we put down 4 as the left-hand digit of the answer.

$$\begin{array}{r} 428 \\ \underline{\times 11} \\ 4\ \_\ \_\ \_ \end{array}$$

Step 2 - Each successive number of the multiplicand is added to its neighbour to the left.

So 4+2 gives 6. This becomes the second digit of the answer.

$$\begin{array}{r} 428 \\ \underline{\times 11} \\ 4\ 6\ \_\ \_ \end{array}$$

Moving to the next two digits, we add 2 + 8 which gives 10.

$$\begin{array}{r} 428 \\ \underline{\times\ 11} \\ 4\ 6\ \_\ \_ \\ 1\ 0 \end{array}$$

Step 3 - If the addition results in two figures, carry over the 1.

The second digit of the sum, which is 0 becomes the third digit of the answer.

$$\begin{array}{r} 428 \\ \underline{\times\ 11} \\ 4\ 6\ 0\ \_ \\ 1\ \uparrow \end{array}$$

Since the addition resulted in two digits, we carry over the one. So 6 now becomes 7.

$$\begin{array}{r} 428 \\ \underline{\times\ 11} \\ 4\ 7\ 0\ \_ \end{array}$$

Step 4 - The last number of the multiplicand becomes the right-hand number of the answer.

So we put down 8 as the last digit of the answer.

$$\begin{array}{r} 428 \\ \underline{\times\ 11} \\ 4\ 7\ 0\ 8 \end{array}$$

The final answer is 4708.

Read this chapter again if necessary. Then go to the practice section and complete the exercises.

You might have understood the technique, but it will take practice before the technique becomes second nature to you.

Students of the Ofpad Mental Math Course **ofpad.com/mathcourse** can post their questions in the discussion section. If you are not a student of the video course, you can email your questions, suggestions & feedback to **hq@ofpad.com**.

If you have enjoyed the book so far, do leave a review on Amazon by visiting **ofpad.com/mathbook**, so that others might also benefit from reading this book.

Once you finished the practice workbook, move on to the next chapter.

Exercises

Download the rich PDFs for these exercises from **ofpad.com/mathexercises**.

1) 54
x 11

2) 34
x 11

3) 46
x 11

4) 984
x 11

5) 723
x 11

6) 342
x 11

7) 424
x 11

8) 216
x 11

9) 923
x 11

10) 3594
x 11

11) 9035
x 11

12) 1593
x 11

13) 6770
x 11

14) 5459
x 11

15) 2696
x 11

16) 7537
x 11

17) 4921
x 11

18) 6871
x 11

Answers

1) 594	7) 4,664	13) 74,470
2) 374	8) 2,376	14) 60,049
3) 506	9) 10,153	15) 29,656
4) 10,824	10) 39,534	16) 82,907
5) 7,953	11) 99,385	17) 54,131
6) 3,762	12) 17,523	18) 75,581

Chapter 5 - The Inefficient Way to Do Math

The technique you learnt in the last chapter is fine for multiplication by 11 but what about larger numbers? We will cover that in the next few chapters. You don't have to memorise special rules for every number. A few generic techniques will help you do arithmetic for any pair of numbers.

Before we proceed to these generic mental math techniques, I must tell you about the real problem we face. This is a mistake you have been making that limits the speed with which you do the math.

And it is not your fault. You have just been taught to do this in school.

Many people think they need a calculator or at least a pen and paper to be able to do the math. The real problem is the fact that you have been taught an inefficient way to do the math in school. And this is what makes it difficult for you to do the math, leave alone doing it in your head.

One of the common lies people believe is that you need an aptitude for math to be good at it. You might have heard that before or in the past you may have even believed it yourself. Yet nothing could be further from the truth.

If you are one of the millions who fell victim to believing this lie, then you must decide now to believe the truth instead. Because if you don't, you will never experience the usefulness of math and numbers in your life, even if you are extremely smart, to begin with.

It is not your fault because you have been led to believe this by our math education system. So if you struggle to do math in your head don't blame yourself, blame the system which still teaches students outdated ways to do math. The system is designed in such a way that only 1% of people get it while the rest of us merely get by.

When I reveal the simple secret of doing mental math in the next chapter, you will be shocked why they never taught you this at school. And when you learn it, math will immediately start becoming more useful and enjoyable.

The truth is that the way we were taught math in school only slows us down because it uses too much of your working memory.

Working memory is the short-term memory you need to organise information to complete a task. It is like the RAM of your computer.

The way we were taught to do math in school is so inefficient that it eats up so much of our working memory when we try to do the math in our head. So our brain slows down like an overloaded computer. This is what makes mental math so hard to do.

To do math faster in your head, you just have to do the opposite of what you have been taught in school. I'll show you exactly how to do this in just a few minutes.

First, I need to show you why you need to avoid doing what you have been taught in school. This might surprise you. Most people think this is the only one way to calculate, but in reality, this is what slows you down.

I will show you what I mean with an example.

Say we wanted to multiply 73210 by 3.

73201
x   3

We will start by multiplying one and three to get three. Then we multiply zero and three to get zero. We proceed to multiply two and three to get six. Then we multiply three and three to get nine, and finally, we multiply seven and three to get twenty-one.

This wasn't very hard, and in fact, it would only take most people seconds to multiply the individual numbers.

However, to get the final answer, you need to remember every single digit you calculated so far to put them back together. You might even end up multiplying again because you forgot one of the numbers.

So it takes quite a bit of effort to get the answer of the multiplication because you spend so much time trying to recall the numbers you already calculated.

Math would be a whole lot easier to do in your head if you didn't have to remember so many numbers.

In school, we have been taught to write down the numbers on paper to free up our working memory. There is another way, and I will show you that in a moment when we cover the LR method.

Now imagine when we tried to multiply 73201 x 3, if you could have come up with the answer, in the time it took you to multiply the individual numbers. Wouldn't you have solved the problem faster than the time it would have taken you to punch in the numbers inside a calculator?

I will show you how to do that in the next chapter.

Chapter 6 - Introducing The LR Method

The secret of doing mental math fast in your head is to do the opposite of what you have been taught in school. This technique is called the LR method.

Let me tell you if you do not do this, any trick you can learn to do mental math will be useless. This secret is like the rocket fuel for your mental math skills.

Everything else just builds on top of this. This simple technique works so well because it frees your working memory almost completely.

LR stands for **L**eft to **R**ight.

The secret of doing mental math is to calculate from left to right instead of from right to left.

This is the opposite of what you have been taught in school.

Let us try to do the earlier example where we multiplied 73201 x 3. But this time let us multiply from left to right.

Try to do it yourself before reading further. I bet you will have no trouble calling out the answer to the multiplication problem.

Multiplying 7 x 3 gives 21.

$$\begin{array}{r} 73201 \\ \underline{\times\qquad 3} \\ 21\_\ \_\ \_\ \_ \end{array}$$

Multiplying 3 x 3 gives 9.

$$\begin{array}{r} 73201 \\ \underline{x \quad\quad 3} \\ 219 \_ \_ \_ \end{array}$$

Multiplying 3 x 2 gives 6.

$$\begin{array}{r} 73201 \\ \underline{x \quad\quad 3} \\ 2196 \_ \_ \end{array}$$

Multiplying 0 x 3 gives 0.

$$\begin{array}{r} 73201 \\ \underline{x \quad\quad 3} \\ 21960 \_ \end{array}$$

Multiplying 3 x 1 gives 3.

$$\begin{array}{r} 73201 \\ \underline{x \quad\quad 3} \\ 219603 \end{array}$$

Did you notice that you started to call out the answer before you even finished the whole multiplication problem? You don't have to remember a thing to recall and use later. So you end up doing math a lot faster.

You just did this for a one digit multiplier 3. Imagine calculating with the same speed for more complex numbers like two and three digit multipliers.

Imagine doing addition, subtraction and division with the same speed.

Let me assure you that it is as ridiculously simple as multiplying using the LR method and this will be covered in detail in the next two chapters.

Before we cover that, walk with me for a moment as we imagine you waking up tomorrow being able to do lightning fast math in your head. Your family and friends are going to look at you like you are some kind of a genius.

Since calculations are done in your head, you would have acquired better memory habits in the process. Your concentration and your ability to think quickly would have also improved. So you will not just look like a genius. You will actually be one.

You know what the best part is? Immediately after completing this book, you will start thinking like a genius. This will, in turn, start to positively influence other areas of your life. I am really excited to show you how.

Chapter 7 - Addition & Subtraction

In this chapter, we will look at how to do addition & subtraction fast using the LR Method. The techniques are similar so we will cover both addition and subtraction in the same chapter.

LR Addition

We saw in the earlier chapter that the secret of mental calculation is to calculate from left to right instead of right to left. When you do this, you will start calling out the answer, before you even complete the full calculation.

Solving math from right to left might be good for pen and paper math. However, when you do math the way you have been taught in school, you will be generating the answer in reverse, and this is what makes math hard to do in your head.

LR method might seem unnatural at first, but you will discover that solving math from left to right, is the most natural way to do calculations in your head. So let us apply this to addition.

Example 1
Add the following numbers together:

$$5\ 3\ 2\ 1$$
$$\underline{+1\ 2\ 3\ 4}$$
$$-\ -\ -\ -$$

The rule is simple. Add from left to right. One digit at a time. So adding 5 + 1 gives 6.

$$5\ 3\ 2\ 1$$
$$\underline{+1\ 2\ 3\ 4}$$
$$6\ \_\ \_\ \_$$

Adding 3 + 2 gives 5.

$$\begin{array}{r} 5\,3\,2\,1 \\ +1\,2\,3\,4 \\ \hline 6\,5\,\_\,\_ \end{array}$$

Adding 2 + 3 gives 5.

$$\begin{array}{r} 5\,3\,2\,1 \\ +1\,2\,3\,4 \\ \hline 6\,5\,5\,\_ \end{array}$$

Adding 1 + 4 gives 5. And you have the answer 6555.

$$\begin{array}{r} 5\,3\,2\,1 \\ +1\,2\,3\,4 \\ \hline 6\,5\,5\,5 \end{array}$$

The individual steps were broken as a way to represent the mental process involved to arrive at the answer. When you do the calculation, it should only take you seconds to arrive at the final answer.

Example 2

Let us try another example. Add the following numbers together:

$$\begin{array}{r} 9\,8\,8\,1 \\ +1\,2\,3\,4 \\ \hline \_\,\_\,\_\,\_ \end{array}$$

The rule is simple. Add from left to right. One digit at a time.

So adding 9 + 1 gives 10.

$$\begin{array}{r} 9\,8\,8\,1 \\ +1\,2\,3\,4 \\ \hline 1\,0\,\_\,\_\,\_ \end{array}$$

Adding 8 + 2 gives 10.

$$
\begin{array}{r}
9\,8\,8\,1 \\
+\,1\,2\,3\,4 \\
\hline
1\,0\,\_\,\_\,\_ \\
1\,0
\end{array}
$$

Carry over the 1, and the 0 becomes 1.

$$
\begin{array}{r}
9\,8\,8\,1 \\
+\,1\,2\,3\,4 \\
\hline
1\,1\,0\,\_\,\_
\end{array}
$$

Adding 8 + 3 gives 11.

$$
\begin{array}{r}
9\,8\,8\,1 \\
+\,1\,2\,3\,4 \\
\hline
1\,1\,0\,\_\,\_ \\
1\,1
\end{array}
$$

Carry over the 1, and the 0 becomes 1.

$$
\begin{array}{r}
9\,8\,8\,1 \\
+\,1\,2\,3\,4 \\
\hline
1\,1\,1\,1\,\_
\end{array}
$$

Adding 1 + 4 gives 5

$$
\begin{array}{r}
9\,8\,8\,1 \\
+\,1\,2\,3\,4 \\
\hline
1\,1\,1\,1\,5
\end{array}
$$

And you have the final answer 11,115.

Example 3

Let us try another example.

Add the following numbers together:

$$8372$$
$$+4636$$
$$-----$$

Add from left to right. One digit at a time. Take a second to apply the technique by yourself. Once you have the answer, check the steps below to see if you got your answer right.

Add from left to right. So adding 8 + 4 gives 12

$$8372$$
$$+4636$$
$$12\_\_\_$$

Adding 3 + 6 gives 9.

$$8372$$
$$+4636$$
$$129\_\_$$

Adding 7 + 3 gives 10.

$$8372$$
$$+4636$$
$$129\_\_$$
$$10$$

Carry over the 1, and the 29 becomes 30.

$$8372$$
$$+4636$$
$$1300\_$$

Adding 2 + 6 gives 8.

$$8372$$
$$+4636$$
$$13008$$

You have the final answer 13,008.

If you got your answer wrong, don't worry. Just revisit the techniques and examples we covered in this chapter.

Tip - When you do the problems in your head, don't just visualise it in your mind, try to hear them as well. For example, when you are adding 8432 + 4636 say eight **thousand** four hundred and thirty-two plus four **thousand** six hundred and thirty-six. You stress the digit you are adding (thousand in this case) to keep track of where you are at.

You can do this for all methods taught in this book. When initially solving the problems, practice the problems by saying it out loud. Saying something out loud will serve as an additional memory aid which clears up more of your working memory.

Rounding Up Before Calculating

A useful technique for addition is to round up the number first, before doing the addition.

$$4529 = 5000 + ?$$

But the problem is finding out how much you rounded up so you can use it later on.

Take the above example. It is easy to round up 4529 to 5000. However, you need to know how much you rounded up. Without a technique to find that out, you will end up doing a subtraction of 5000 − 4529 which will defeat the whole purpose of rounding up.

Finding How Much You Rounded

To find how much you rounded up without explicitly doing the subtraction, remember the following:

Step 1 - The last digits add up to 10.

Step 2 - The remaining digits add up to 9.

Let us look at our earlier example:

$$4529 = 5000 + 471$$

In this example, when you round up 4529 to 5000, you have rounded up by 471.

The last digits of 4529 and 471 add up to 10.

$$452\underline{9} = 5000 + 47\underline{1}$$
$$9 + 1 = 10$$

All the other digits 5 and 2 (in 4529) & 4 and 7 (471) add up to 9.

$$4\underline{52}9 = 5000 + \underline{47}1$$
$$5 + 4 = 9$$
$$2 + 7 = 9$$

So 5 + 4 is 9 & 2 + 7 is again 9.

Knowing that the last digits add up to 10 and other digits add up to 9, you will be able to arrive at the amount you rounded up quickly.

Let us do a few rounding up exercises.

Exercise 1

Round up 84791 to 100,000 and find out how much you rounded up by.

$$84791 = 100,000 + ?$$

Remember the last digits add up to 10. All the other digits add up to 9.

Take a second to call out the answer.

Then proceed below to check if you got it right.

The answer is 15209.

9 in 15209 adds up with 1 of 84791 to give 10.

All the other digits add up to 9.

Exercise 2
Let us try another number.

Round up 7423 to 10,000 and find out how much you rounded up by.

$$7423 = 10,000 + ?$$

Remember the last digits add up to 10. All the other digits add up to 9.

Take a second to call out the answer.

Then proceed below to check if you got it right.

The answer is 2577.

7 in 2577 adds up with 3 of 7423 to give 10.

All the other digits add up to 9.

Exercise 3

Let us try another number.

Round up 892 to 1000 and find out how much you rounded up by.

$$892 = 1000 + ?$$

Take a second to call out the answer.

Then proceed below to check if you got it right.

The answer is 108.

8 in 108 adds up with 2 of 892 to give 10.

All the other digits add up to 9.

Exercise 4

Let us try one last example.

Round up 27 to 100 and find out how much you rounded up by.

$$27 = 100 + ?$$

Remember the last digits add up to 10. All the other digits add up to 9.

Take a second to call out the answer.

Then proceed below to check if you got it right.

The answer is 73.

3 in 73 adds up with 7 of 27 to give 10.

The other digits, i.e. 2 in 27 and 7 in 73 add up to 9.

LR Addition With Rounding Up

Now that you learnt how to find out how much you rounded up, let us look at the method to do LR Addition with rounding up. It is extremely simple.

Step 1 - Round up a number.

Step 2 - Add from left to right to the amount to which you rounded up.

Step 3 - Subtract the amount you rounded up from the sum.

Let us look at an example and try to do left to right addition after rounding up one of the numbers.

Example 1
Add the following numbers:

$$9981$$
$$+\ 1234$$
$$----$$

Step 1 - Round up the first number.

$$9981 = (10,000 - 19)$$
$$+\ 1234$$
$$----$$

So when you round up 9981, you have 10,000, and the amount you rounded up is 19.

$$10,000 - 19$$
$$+\ 1234$$
$$----$$

If you have trouble finding the amount you rounded up, check the section of rounding up we covered where we saw how the last digits add up to 10 and the remaining digits add up to 9.

Step 2 - Now add from left to right to the amount you rounded up.

$$10,000 - 19$$
$$+ \ 1234$$
$$11,234$$

So if you add 10,000 with 1234, you get 11234. Rounding up made the addition process easy.

Step 3 - Now subtract the amount you rounded up from the sum.

$$11,234$$
$$- \qquad 19$$
$$-\ -\ -\ -\ -$$

So subtract 19 from 11234.

Subtract from left to right.

We can put down 112 before proceeding to do the subtraction.

$$11,234$$
$$- \qquad 19$$
$$1\ 1\ 2\ \_\ \_$$

Subtract 3 - 1 is 2.

$$11,234$$
$$- \qquad 19$$
$$1\ 1\ 2\ 2\ \_$$

Next, you have to subtract 4 − 9 so you have to borrow 1 from the 2. So now the 2 becomes 1.

$$
\begin{array}{r}
11{,}234 \\
-19 \\
\hline
1\,1\,2\,1\,\_ \\
\end{array}
$$

And 14 − 9 is 5.

$$
\begin{array}{r}
11{,}234 \\
-19 \\
\hline
11{,}215 \\
\end{array}
$$

That is your answer 11,215.

Note that we deep dived into the steps so that you have clarity and understanding. When you do the steps in your mind, it will take less than 5 seconds to do this entire calculation. Each chapter will explain the technique you should apply, and the practice exercises will take the speed with which you apply the technique to the next level.

Example 2
Let us try another example now.

Add 5492 with 8739.

$$
\begin{array}{r}
5492 \\
+\ 8739 \\
\hline
-\ -\ -\ - \\
\end{array}
$$

Step 1 - Round up the first number.

$$
\begin{array}{r}
5492 \quad (5500 - 8) \\
+\ 8739 \\
\hline
-\ -\ -\ - \\
\end{array}
$$

So when you round up 5492, you have 5500, and the amount you rounded up is 8. Note that I could just round up to 6000 with the rounded up value being 508. But rounding up to 5500 will simplify the subtraction step in this example. The smaller the amount you round up, the easier it is to calculate.

$$\begin{array}{r} 5500 - 8 \\ + \ 8739 \\ \hline \text{---} \end{array}$$

If you have trouble finding the amount you rounded up, check the section of rounding up where we saw how the last digits add up to 10 and the remaining digits add up to 9.

Step 2 - Add from left to right to the amount you rounded up.

So if you add 5500 with 8739, you get 14239. Rounding up made the addition process a little easier.

$$\begin{array}{r} 5500 - 8 \\ + \ 8739 \\ \hline 14{,}239 \end{array}$$

Step 3 - Subtract the amount you rounded up from the sum.

So subtract 8 from 14239. Subtract from left to right.

$$\begin{array}{r} 14{,}239 \\ - \quad 8 \\ \hline 14{,}231 \end{array}$$

Now you have the final answer 14,231.

Example 3

Let us try another example now.

Add 7997 with 5347.

$$7997$$
$$+ \ 5347$$
$$\overline{- \ - \ - \ -}$$

Take a second to apply the technique by yourself as fast as you can. Once you have the answer, you can check the steps below to see if you got your answer right.

Step 1 - Round up the first number.

$$7997 = (8000 - 3)$$
$$+ \ 5347$$
$$\overline{- \ - \ - \ -}$$

So when you round up 7997, you have 8,000, and the amount you rounded up is 3.

$$8000 - 3$$
$$+ \ 5347$$
$$\overline{- \ - \ - \ -}$$

Step 2 - Add from left to right to the amount you rounded up.

So if you add 8,000 with 5347, you get 13347. Rounding up made the addition process easy.

$$8000 - 3$$
$$+ \ 5347$$
$$\overline{13,347}$$

Step 3 - Subtract the amount you rounded up from the sum.

So subtract 3 from 13347.

$$13,347$$
$$- \qquad 3$$
$$\overline{13,344}$$

Now you have the final answer 13344.

If you got your answer wrong, don't worry. Just revisit the techniques and examples we covered in this chapter.

Students of the video course **ofpad.com/mathcourse** can post their questions in the discussion section.

When Should You Round Up For Addition?

Now you understand how to add and round up, should you always be rounding up before adding or should you do it only during specific situations? Sometimes rounding up simplifies the addition whereas during other times, rounding up just complicates the addition and creates an unnecessary step. So how do you decide when to round up and when not to round up? The short answer is that **rounding up should only be done when you have to carry over** a lot of numbers.

Let us try to add 9898 + 4343:

$$
\begin{array}{r}
\mathbf{9898} \\
+\ \mathbf{4343} \\
\hline

\end{array}
$$

You could do this problem with just the LR method without any rounding up. However, if you do that you will have to carry over a number during every step. If you round up the first number, it will make things a lot easier.

You can round up 9898 by 102 to get 10,000.

$$
\begin{array}{r}
\mathbf{9898 = (10000 - 102)} \\
\mathbf{+4343} \\
\hline

\end{array}
$$

Adding 10,000 + 4343 = 14,343.

$$\begin{array}{r} 10000 - 102 \\ +\ 4343 \\ \hline 14{,}343 \end{array}$$

Then subtract the amount you rounded up which is 102 to get 14,241.

$$\begin{array}{r} 14{,}343 \\ -\ \ \ \ 102 \\ \hline 14{,}241 \end{array}$$

Try to do add 9898 + 4343 in your head again, with and without rounding up and you will realize you are flexing fewer neurons when you round up.

So this means rounding up is the magic you do before you do every addition right? No.

Let us say you want to add 4343 + 1234.

$$\begin{array}{r} 4343 \\ +\ 1234 \\ \hline ----\end{array}$$

You don't have to carry over any number to do this calculation. So the LR addition WITHOUT rounding up makes sense. But for the fun of it, try to do the same problem by rounding up. You will find that rounding up only increases the mental effort required to do the math.

So remember to only round up when you have to carry over a lot of numbers. If you don't have to carry over any number or if you have to carry over just one digit, you will end up complicating the problem by rounding up.

Addition Exercises

Download the rich PDFs for these exercises from **ofpad.com/mathexercises**.

01) 33 +20	07) 115 +596	13) 2771 +3216
02) 77 +97	08) 485 +327	14) 6526 +3057
03) 82 +63	09) 114 +164	15) 6491 +2273
04) 157 +836	10) 4942 +2332	16) 3878 +5483
05) 214 +155	11) 7241 +9508	17) 7682 +8903
06) 865 +467	12) 8699 +9897	18) 9616 +9202

Addition Answers

01) 53	07) 711	13) 5,987
02) 174	08) 812	14) 9,583
03) 145	09) 278	15) 8,764
04) 993	10) 7,274	16) 9,361
05) 369	11) 16,749	17) 16,585
06) 1,332	12) 18,596	18) 18,818

LR Subtraction

Now we will look at subtraction. The steps are same as that of addition. We briefly saw how to subtract from left to right when we covered rounding up for addition. So we will just jump straight into an example for subtraction.

Example 1

Subtract 1234 from 5389.

$$\begin{array}{r} 5389 \\ - \underline{1234} \\ \_\ \_\ \_\ \_ \end{array}$$

The rule is simple. Subtract from left to right. One digit at a time. So 5 - 1 gives 4.

$$\begin{array}{r} 5389 \\ - \underline{1234} \\ 4\ \_\ \_\ \_ \end{array}$$

Subtracting 3 - 2 gives 1.

$$\begin{array}{r} 5389 \\ - \underline{1234} \\ 4\ 1\ \_\ \_ \end{array}$$

Subtracting 8 - 3 gives 5.

$$\begin{array}{r} 5389 \\ - \underline{1234} \\ 4\ 1\ 5\ \_ \end{array}$$

Subtracting 9 - 4 gives 5.

$$\begin{array}{r} 5389 \\ - \underline{1234} \\ 4\ 1\ 5\ 5 \end{array}$$

And you have the answer 4155.

Example 2

Let us try another example. Subtract 5741 from 8431.

$$
\begin{array}{r}
\mathbf{8431} \\
\mathbf{-\ 5741} \\
\hline
\_\ \_\ \_\ \_
\end{array}
$$

The rule is simple. Subtract from left to right. One digit at a time.

So 8 - 5 gives 3.

$$
\begin{array}{r}
\mathbf{8431} \\
\mathbf{-\ 5741} \\
\hline
\mathbf{3}\ \_\ \_\ \_
\end{array}
$$

To subtract 4 - 7 you will have to borrow 1 from 3. So the 3 becomes 2.

$$
\begin{array}{r}
\mathbf{8431} \\
\mathbf{-\ 5741} \\
\hline
\mathbf{2}\ \_\ \_\ \_
\end{array}
$$

Subtracting 14 - 7 now gives 7.

$$
\begin{array}{r}
\mathbf{8431} \\
\mathbf{-\ 5741} \\
\hline
\mathbf{2\ 7}\ \_\ \_
\end{array}
$$

To subtract 3 - 4 you will have to borrow 1 from 7. So the 7 becomes 6.

$$
\begin{array}{r}
\mathbf{8431} \\
\mathbf{-\ 5741} \\
\hline
\mathbf{2\ 6}\ \_\ \_
\end{array}
$$

Subtracting 13 - 4 now gives 9.

$$\begin{array}{r} \mathbf{8431} \\ \mathbf{-\ 5741} \\ \hline \mathbf{2\ 6\ 9\ \_} \end{array}$$

Subtracting 1 - 1 gives 0.

$$\begin{array}{r} \mathbf{8431} \\ \mathbf{-\ 5741} \\ \hline \mathbf{2\ 6\ 9\ 0} \end{array}$$

And you have the answer 2690.

Example 3

Let us try another example. Subtract 3756 from 7328.

$$\begin{array}{r} \mathbf{7328} \\ \mathbf{-\ 3756} \\ \hline \mathbf{\_\ \_\ \_\ \_} \end{array}$$

Subtract from left to right. One digit at a time.

Take a second to apply the technique by yourself as fast as you can. Once you have the answer, you can check the steps below to see if you got your answer right.

Subtract from left to right. One digit at a time. So 7 - 3 gives 4.

$$\begin{array}{r} \mathbf{7328} \\ \mathbf{-\ 3756} \\ \hline \mathbf{4\ \_\ \_\ \_} \end{array}$$

To subtract 3 - 7 you will have to borrow 1 from 4. So the 4 becomes 3.

$$\begin{array}{r} 7328 \\ -\ 3756 \\ \hline 3\ \_\ \_\ \_ \end{array}$$

Subtracting 13 - 7 now gives 6.

$$\begin{array}{r} 7328 \\ -\ 3756 \\ \hline 3\ 6\ \_\ \_ \end{array}$$

To subtract 2 - 5 you will have to borrow 1 from 6. So the 6 becomes 5.

$$\begin{array}{r} 7328 \\ -\ 3756 \\ \hline 3\ 5\ \_\ \_ \end{array}$$

Subtracting 12 - 5 now gives 7.

$$\begin{array}{r} 7328 \\ -\ 3756 \\ \hline 3\ 5\ 7\ \_ \end{array}$$

Subtracting 8 - 6 gives 2.

$$\begin{array}{r} 7328 \\ -\ 3756 \\ \hline 3\ 5\ 7\ 2 \end{array}$$

Now you have the final answer 3572.

If you got your answer wrong, don't worry. Just revisit the techniques and examples we covered in this chapter.

Students of the Ofpad Mental Math Course **ofpad.com/mathcourse** can post their questions in the discussion section. If you are not a student of the video course, you can email your questions, suggestions & feedback to **hq@ofpad.com**.

The last problem was a little harder than the rest because you had to borrow numbers from the neighbour during two separate steps. That is where the magic of rounding up will come in.

LR Subtraction With Rounding Up

When you did LR Subtraction, you might have noticed that it was a bit tedious to do the subtraction when you had to borrow numbers from the neighbour. This is where rounding up numbers really helps. Rounding up also applies to subtraction. But instead of subtracting the amount you rounded up (as we did in addition), you will now add the amount you rounded up.

Rounding up before calculating is usually more useful for subtraction than it is for addition. That is because for most of us it is generally easier to do mental addition than it is to do mental subtraction. Borrowing numbers from the neighbour during subtraction is not as easy as carrying over numbers during addition. When you round up before you subtract, you make the subtraction problem into a simple addition problem.

If required, revisit the section of rounding up where we covered how to find how much you rounded up. The rule to round up and subtract is simple:

Step 1 - Round up the second number.

Step 2 - Subtract from left to right the amount you rounded up to.

Step 3 - Add the amount you rounded up with the subtracted amount.

Let us look at an example.

Example 1

Let us try to do left to right subtraction after rounding up the numbers. Subtract 3898 from 4530.

$$\begin{array}{r} 4530 \\ - \underline{3898} \\ \hline ----\end{array}$$

Notice that this is a hard subtraction problem to do with the straightforward LR subtraction method because you have to borrow a digit as you calculate almost every number. However, this problem is a piece of cake when you round up before you calculate. I will show you how.

Step 1 - Round up the second number.

$$\begin{array}{r} 4530 \\ - \underline{3898} \end{array} = - (4000 - 102)$$
$$----$$

So when you round up 3898, you have 4,000, and the amount you rounded up is 102.

$$\begin{array}{r} 4530 \\ - \underline{4000} + 102 \\ \hline ----\end{array}$$

If you have trouble finding the amount you rounded up, check the section of rounding up in this chapter where we saw how the last digits add up to 10 and the remaining digits add up to 9.

Step 2 - Now subtract from left to right the amount you rounded up to.

So if you subtract 4000 from 4530, you get 530. Rounding up made the subtraction process easy.

$$\begin{array}{r} 4530 \\ \underline{- \textbf{4000}} + 102 \\ 530 \end{array}$$

Step 3 - Now add the amount you rounded up with the subtracted amount.

So add 102 to 530. Add from left to right.

$$\begin{array}{r} 530 \\ \underline{+102} \\ 632 \end{array}$$

You get 632 which is your final answer. Note that we deep dived into the steps so that you have clarity and understanding. When you do the steps in your mind, it should take less than 5 seconds to do this entire calculation.

Example 2

Let us try another example. Subtract 4998 from 7520.

$$\begin{array}{r} 7520 \\ \underline{- 4998} \\ ---- \end{array}$$

Again doing LR subtraction without rounding up would mean you have to borrow so many numbers as you calculate. So rounding up will simplify the mental process.

Step 1 - Round up the second number.

$$
\begin{array}{r}
7520 \\
\underline{-\ 4998} \\

\end{array} = -(5000 - 2)
$$

So when you round up 4998, you have 5,000, and the amount you rounded up is 2.

$$
\begin{array}{r}
7520 \\
\underline{-\ 5000} + 2 \\

\end{array}
$$

Step 2 - Subtract from left to right the amount you rounded up to.

So if you subtract 5000 from 7520, you get 2520. Rounding up made the subtraction process easy.

$$
\begin{array}{r}
7520 \\
\underline{-\ 5000} + 2 \\
2520
\end{array}
$$

Step 3 - Now add the amount you rounded up with the subtracted amount.

So add 2 to 2520.

$$
\begin{array}{r}
2520 \\
\underline{+\quad 2} \\
2522
\end{array}
$$

You get 2522 which is your final answer.

Example 3

Let us try another example. Subtract 2796 from 8734.

$$
\begin{array}{r}
8734 \\
\underline{-\ 2796} \\

\end{array}
$$

Remember to round up the second number and then do the subtraction.

Take a second to apply the technique by yourself as fast as you can. Once you have the answer, you can check the steps below to see if you got your answer right.

Step 1 - Round up the second number.

$$\begin{array}{r} 8734 \\ \underline{- 2796} = - (3000 - 204) \\ \underline{- \; - \; - \; -} \end{array}$$

So when you round up 2796, you have 3,000, and the amount you rounded up is 204.

$$\begin{array}{r} 8734 \\ \underline{- 3000} + 204 \\ \underline{- \; - \; - \; -} \end{array}$$

If you have trouble finding the amount you rounded up, check the section of rounding up in this chapter where we saw how the last digits add up to 10 and the remaining digits add up to 9.

Step 2 - Now subtract from left to right the amount you rounded up to.

So if you subtract 3000 from 8734, you get 5734. Rounding up made the subtraction process easy.

$$\begin{array}{r} 8734 \\ \underline{- 3000} + 204 \\ 5734 \end{array}$$

Step 3 - Add the amount you rounded up with the subtracted amount.

So add 204 to 5734. Add from left to right.

5734
+204
5938

You get 5938 which is your final answer.

If you got your answer wrong, don't worry. Just revisit the techniques and examples we covered in this chapter.

Students of the Ofpad Mental Math Course **ofpad.com/mathcourse** can post their questions in the discussion section. If you are not a student of the video course, you can email your questions, suggestions & feedback to **hq@ofpad.com**.

When Should You Round Up For Subtraction?

Just like addition, should you always be rounding up before subtracting or should you do it only during specific situations? Sometimes rounding up simplifies the subtraction whereas during other times, rounding up just complicates the subtraction and creates an unnecessary step. So how do you decide when to round up and when not to round up?

The short answer is that **rounding up should only be done when you have to borrow** a lot of numbers for subtraction just like how you had to carry over a lot of numbers for addition.

Rounding up is more valuable for subtraction than it is for addition because borrowing numbers in your head during subtraction is a lot harder to do compared to carrying over numbers that you do in addition.

In subtraction you round up the number when you have to borrow a lot of numbers. Let us take for example 53,441 - 49,898.

$$\begin{array}{r} \mathbf{53,441} \\ \underline{\mathbf{- 49,898}} \\ \mathbf{- - - -} \end{array}$$

In this subtraction, except for the first numbers (5 - 4) you have to borrow a number during every step. So it makes sense to round up the second number before doing the subtraction.

Rounding up the second number 49,898 by 102 we get 50,000.

$$\begin{array}{r} \mathbf{53,441} \\ \underline{\mathbf{- 49,898}} = \mathbf{- (50,000 - 102)} \\ \mathbf{- - - -} \end{array}$$

Subtracting 53,441 - 50,000 we get 3441.

$$\begin{array}{r} \mathbf{53,441} \\ \underline{\mathbf{- 50,000}} + \mathbf{102} \\ \mathbf{3,441} \end{array}$$

Adding 102 we get 3,543.

$$\begin{array}{r} \mathbf{3,441} \\ \underline{\mathbf{+102}} \\ \mathbf{3,543} \end{array}$$

Now try doing the same subtraction 53,441 - 49,898 without rounding up. You might find it more strenuous to do the subtraction without round up.

Do you round up before doing every subtraction? Not really.

Subtract 9889 - 4343.

$$\begin{array}{r} \mathbf{9889} \\ \mathbf{-\ 4343} \\ \hline \_\ \_\ \_\ \_ \end{array}$$

You can do the straight forward LR subtraction here to get the answer. You can do rounding up and solve the math but it will only result in an unnecessary step.

So remember to only round up when you have to borrow numbers. If you don't have to borrow any number you will be better off doing the subtraction without rounding up.

Read this chapter again if necessary, and then go to the practice section and complete the exercises.

You might have understood the technique, but it will take practice before the technique becomes second nature to you.

If you have enjoyed the book so far, do leave a review on Amazon by visiting **ofpad.com/mathbook**, so that others might also benefit from reading this book.

Subtraction Exercises

Download the rich PDFs for these exercises from **ofpad.com/mathexercises**.

01) 77
  - 55
	07) 867
	  - 798
		13) 8450
		  - 8109

02) 81
  - 61
	08) 709
	  - 202
		14) 9880
		  - 2941

03) 54
  - 25
	09) 905
	  - 227
		15) 8827
		  - 8718

04) 758
  - 482
	10) 9125
	  - 5305
		16) 5508
		  - 2741

05) 740
  - 424
	11) 5487
	  - 1120
		17) 6991
		  - 1399

06) 919
  - 872
	12) 8017
	  - 7676
		18) 7610
		  - 3171

Subtraction Answers

01) 16	07) 69	13) 341
02) 20	08) 507	14) 6,939
03) 29	09) 678	15) 109
04) 276	10) 3,820	16) 2,767
05) 316	11) 4,377	17) 5,592
06) 47	12) 341	18) 4,439

Chapter 8 - LR Multiplication

In this chapter, we will look at how to multiply any set of numbers fast using the LR Method.

As we saw earlier, LR Stands for Left to Right. You have already seen the power of calculating from left to right in the previous chapters for addition and subtraction. In this chapter, we will apply the same concept to multiplication.

The secret of mental multiplication like addition and subtraction is multiply from left to right instead of right to left.

The beauty of this method is that it is a generic technique that can be applied to any pairs of numbers.

There is one pre-requisite for this chapter. You need to know your multiplication tables. You should memorize your multiplication table if you haven't done so already.

Below is a quick refresher if you need it:

X	1	2	3	4	5	6	7	8	9
1	1	2	3	4	5	6	7	8	9
2	2	4	6	8	10	12	14	16	18
3	3	6	9	12	15	18	21	24	27
4	4	8	12	16	20	24	28	32	36
5	5	10	15	20	25	30	35	40	45
6	6	12	18	24	30	36	42	48	54
7	7	14	21	28	35	42	49	56	63
8	8	16	24	32	40	48	56	64	72
9	9	18	27	36	45	54	63	72	81

When you look at the examples in this chapter and later do the practice, learn to visualise the numbers in your mind.

That is where practice comes in. The more you practice, the better memory habits you will develop, which will make remembering numbers easier.

One Digit Multiplier

We will start off by applying LR Multiplication to one digit multipliers.

Example 1

Let us try an example. Multiply 5321 by 4.

$$\begin{array}{r} 5321 \\ \times\quad 4 \\ \hline \text{\_ \_ \_ \_} \end{array}$$

Multiply from left to right. Multiplying 5 x 4 = 20.

$$\begin{array}{r} 5321 \\ \times\quad 4 \\ \hline 20\text{\_ \_ \_} \end{array}$$

Multiplying 3 x 4 = 12.

$$\begin{array}{r} 5321 \\ \times\quad 4 \\ \hline 212\text{\_ \_} \end{array}$$

Multiplying 2 x 4 = 8.

$$\begin{array}{r} 5321 \\ \times\quad 4 \\ \hline 2128\text{\_} \end{array}$$

Multiplying 1 x 4 = 4.

$$\begin{array}{r} 5321 \\ \times\quad 4 \\ \hline 21,284 \end{array}$$

Now you have your final answer 21,284.

Example 2

Let us try another example. Multiply 7142 by 6.

$$\begin{array}{r} \mathbf{7142} \\ \underline{\mathbf{x \quad 6}} \\ \mathbf{-\ -\ -\ -} \end{array}$$

Multiply from left to right. When doing this calculation, recall how we covered carrying numbers over when we multiplied by 11. The same applies here and in other techniques in this book.

Multiply 7 x 6 = 42.

$$\begin{array}{r} \mathbf{7142} \\ \underline{\mathbf{x \quad 6}} \\ \mathbf{42\ \_\ \_\ \_} \end{array}$$

Multiplying 1 x 6 = 6.

$$\begin{array}{r} \mathbf{7142} \\ \underline{\mathbf{x \quad 6}} \\ \mathbf{426\ \_\ \_} \end{array}$$

Multiplying 4 x 6 = 24.

$$\begin{array}{r} \mathbf{7142} \\ \underline{\mathbf{x \quad 6}} \\ \mathbf{426\ \_\ \_} \\ \mathbf{24} \end{array}$$

Carry over the 2, so the 6 becomes 8.

$$\begin{array}{r} \mathbf{7142} \\ \underline{\mathbf{x \quad 6}} \\ \mathbf{4284\ \_} \end{array}$$

Multiply 2 x 6 = 12.

$$
\begin{array}{r}
7142 \\
\times\quad 6 \\
\hline
4284\ \_ \\
12
\end{array}
$$

Carry over the 1, so the 4 becomes 5.

$$
\begin{array}{r}
7142 \\
\times\quad 6 \\
\hline
42852
\end{array}
$$

And you have the answer 42,852.

When you have to carry over numbers, the difficult part may be holding the numbers you already calculated in your memory. Saying the numbers out loud will help free up some of the memory. However, with practice, you will be able to effortlessly remember even without doing this, because practice will improve your concentration and memory. When you are doing the examples in this book, you might be tempted to glance at the problem to remind yourself. This is fine but try to hold the numbers in your head. It will become easier the more you practice doing it.

Example 3
Let us try another example.

Multiply 9432 by 3.

$$
\begin{array}{r}
9432 \\
\times\quad 3 \\
\hline
\text{-- -- -- --}
\end{array}
$$

Multiply from left to right.

Take a second to apply the technique by yourself as fast as you can. Once you have the answer, you can check the steps below to see if you got your answer right.

Multiplying 9 x 3 = 27.

$$\begin{array}{r} 9432 \\ \underline{\times\quad 3} \\ 27\_\_\_ \end{array}$$

Multiplying 4 x 3 = 12.

$$\begin{array}{r} 9432 \\ \underline{\times\quad 3} \\ 27\_\_\_ \\ 12 \end{array}$$

Carry over the 1, so the 7 becomes 8.

$$\begin{array}{r} 9432 \\ \underline{\times\quad 3} \\ 282\_\_ \end{array}$$

Multiplying 3 x 3 = 9.

$$\begin{array}{r} 9432 \\ \underline{\times\quad 3} \\ 2829\_ \end{array}$$

Multiplying 2 x 3 = 6.

$$\begin{array}{r} 9432 \\ \underline{\times\quad 3} \\ 28{,}296 \end{array}$$

The final answer is 28,296.

If you got your answer wrong, don't worry. Just revisit the techniques and examples we covered in this chapter.

Students of the Ofpad Mental Math Course **ofpad.com/mathcourse** can post their questions in the discussion section. If you are not a student of the video course, you can email your questions, suggestions & feedback to **hq@ofpad.com**.

Practice is necessary for you to master these techniques. At first, mental math might seem like a thing only a genius can do. However, it is like learning how to ride a bicycle - you cannot forget it, once you have learnt it.

Exercises

Download the rich PDFs for these exercises from **ofpad.com/mathexercises**.

01) 86 x 5	07) 168 x 5	13) 8013 x 6
02) 45 x 7	08) 906 x 4	14) 5088 x 9
03) 60 x 9	09) 520 x 7	15) 7895 x 9
04) 510 x 7	10) 1816 x 4	16) 2766 x 3
05) 398 x 6	11) 3619 x 7	17) 5781 x 5
06) 645 x 9	12) 8824 x 5	18) 5451 x 7

Answers

01) 430	07) 840	13) 48,078
02) 315	08) 3,624	14) 45,792
03) 540	09) 3,640	15) 71,055
04) 3,570	10) 7,264	16) 8,298
05) 2,388	11) 25,333	17) 28,905
06) 5,805	12) 44,120	18) 38,157

One Digit Multiplier With Rounding Up

Just like in LR addition and subtraction, rounding up and multiplying is especially useful when the numbers end in 7, 8 or 9.

Method

The method is simple:

Step 1 - Round up the number.

Step 2 - Multiply from left to right.

Step 3 - Multiply the amount you rounded up.

Step 4 - Subtract the numbers from the previous two steps.

Let us look at an example to illustrate the method.

Example 1

Multiply 68 by 3.

$$\begin{array}{r} \mathbf{68} \\ \underline{\mathbf{x\ 3}} \\ \_\ \_\ \_ \end{array}$$

Step 1 - Round up the number.

$$\begin{array}{r} \mathbf{68\ (70-2)} \\ \underline{\mathbf{x\ 3}} \\ \_\ \_\ \_ \end{array}$$

So round up 68 to 70. You have rounded up by 2.

$$\begin{array}{r} \mathbf{70-2} \\ \underline{\mathbf{x\ 3}} \\ \_\ \_\ \_ \end{array}$$

Step 2 - Multiply from left to right.

Multiply 70 by 3, and you have 210.

$$
\begin{array}{r}
70 - 2 \\
\underline{\times\, 3} \\
210
\end{array}
$$

Step 3 - Multiply the amount you rounded up.

Multiplying 2 x 3 = 6.

$$
\begin{array}{rr}
70 & - & 2 \\
\underline{\times\, 3} & & \underline{\times\, 3} \\
210 & & 6
\end{array}
$$

Step 4 - Subtract the numbers from the previous two steps.

Subtracting 210 – 6 we get 204.

$$
\begin{array}{rr}
70 & - & 2 \\
\underline{\times\, 3} & & \underline{\times 3} \\
210 & - & 6 = 204
\end{array}
$$

And now you have the answer as 204.

$$
\begin{array}{r}
68 \\
\underline{\times\, 3} \\
204
\end{array}
$$

The entire multiplication process has been greatly simplified because you rounded up.

Example 2

Let us try another example. Multiply 96 by 7.

$$
\begin{array}{r}
96 \\
\underline{\times\, 7} \\
- - -
\end{array}
$$

Step 1 - Round up the number.

$$96 = 100 - 4$$
$$\underline{\times\,7}$$
$$-\ -\ -$$

So we round up 96 to 100. We have rounded up by 4.

Step 2 - Multiply from left to right.

$$100 - 4$$
$$\underline{\times\,7}$$
$$-\ -\ -$$

Multiply 100 by 7, and you have 700.

$$100 - 4$$
$$\underline{\times\,7}$$
$$700$$

Step 3 - Multiply the amount you rounded up.

Multiplying 4 x 7 = 28.

$$
\begin{array}{ll}
100 & -\quad 4 \\
\underline{\times\,7} & \underline{\times 7} \\
700 & \quad 28
\end{array}
$$

Step 4 - Subtract the numbers from the previous two steps.

Subtracting 700 – 28 we get 672.

$$
\begin{array}{ll}
100 & -\quad 4 \\
\underline{\times\,7} & \underline{\times\,7} \\
700 & -\ \ 28 = 672
\end{array}
$$

And now you have the answer as 672.

If you have trouble in this last step of subtraction, remember the rule for rounding up we saw in the previous chapter. The same rule can be applied to do the subtraction quickly. Notice the last digits (8 in 28 and 2 in 672) add up to 10 and the other digits (2 in 28 and 7 and 672) add up to 9. The rounding up rule can be applied to subtraction of numbers like this, and you can find out the answer without explicitly doing the subtraction.

A seemingly complicated multiplication problem has become so simple because we just rounded up.

Example 3

Let us try another example.

Multiply 398 by 9.

$$\begin{array}{r} \textbf{398} \\ \underline{\textbf{x 9}} \\ \text{— — —} \end{array}$$

Take a second to apply the technique by yourself as fast as you can. Once you have the answer, you can check the steps below to see if you got your answer right.

Step 1 - Round up the number.

$$\begin{array}{r} \textbf{398 = (400 – 2)} \\ \underline{\textbf{x 9}} \\ \text{— — —} \end{array}$$

So we round up 398 to 400. We have rounded up by 2.

$$\begin{array}{r} \textbf{400 – 2} \\ \underline{\textbf{x 9}} \\ \text{— — —} \end{array}$$

Step 2 - Multiply from left to right.

Multiply 400 by 9, and you have 3600.

$$400 - 2$$
$$\underline{\times\,9}$$
$$3600$$

Step 3 - Multiply the amount you rounded up.

Multiplying 2 x 9 = 18.

$$\begin{array}{cc} 400 & - & 2 \\ \underline{\times\,9} & & \underline{\times 9} \\ 3600 & & 18 \end{array}$$

Step 4 - Subtract the numbers from the previous two steps.

Subtract 3600 – 18.

$$\begin{array}{cc} 400 & - & 2 \\ \underline{\times\,9} & & \underline{\times\,9} \\ 3600 & - & 18 \end{array}$$

And now you have the answer as 3582.

$$\begin{array}{cc} 400 & - & 2 \\ \underline{\times\,9} & & \underline{\times 9} \\ 3600 & - & 18 = 3582 \end{array}$$

If you have trouble in this last step of subtraction, remember the rule we saw in the previous chapter where the last digits (8 in 18 and 2 in 3592) add up to 10 and the other digits (1 in 18 and 8 and 3582) add up to 9. It will greatly simplify the last step, and you will be able to find the answer without explicitly doing any subtraction.

If you got your answer wrong, don't worry. Just revisit the techniques and examples we covered in this chapter.

You can post your questions in the discussion section of **ofpad.com/mathcourse** or email me at **hq@ofpad.com**.

Exercises

Download the rich PDFs for these exercises from **ofpad.com/mathexercises**.

01) 57 x 8	07) 498 x 3	13) 4998 x 6
02) 28 x 3	08) 298 x 8	14) 1999 x 7
03) 67 x 8	09) 197 x 7	15) 8999 x 3
04) 798 x 3	10) 4997 x 4	16) 6997 x 9
05) 798 x 8	11) 6998 x 5	17) 4997 x 6
06) 298 x 9	12) 2998 x 4	18) 6998 x 8

Answers

01) 456	07) 1,494	13) 29,988
02) 84	08) 2,384	14) 13,993
03) 536	09) 1,379	15) 26,997
04) 2,394	10) 19,988	16) 62,973
05) 6,384	11) 34,990	17) 29,982
06) 2,682	12) 11,992	18) 55,984

LR Multiplication – Two Digit Multiplier

We looked at multiplication using a 1 digit multiplier. Let us look at a 2 digit multiplier now. To multiply by a 2 digit multiplier.

Method

Step 1 - Break the multiplicand.

Step 2 - Multiply from left to right.

Step 3 - Add the individual answers together.

Let us look at an example.

Example 1

So let us multiply 36 by 32.

$$\begin{array}{r} 36 \\ \underline{\times\ 32} \\ - - - - \end{array}$$

Step 1 - Break the multiplier.

32 is broken down as 30 + 2.

$$\begin{array}{r} 36 \\ \underline{\times\ 32} = (30 + 2) \\ - - - - \end{array}$$

Step 2 - Multiply from left to right.

Multiplying 36 x 30 = 1080.

$$\begin{array}{r} 36 \\ \underline{\times\ 30} + 2 \\ 1080 \end{array}$$

Multiplying 36 x 2 = 72.

$$\begin{array}{cc} 36 & 36 \\ \underline{\text{x } 30} + & \underline{\text{x } 2} \\ 1080 & 72 \end{array}$$

Step 3 - Add the individual answers together.

Adding 1080 + 72 = 1152.

$$\begin{array}{cc} 36 & 36 \\ \underline{\text{x } 30} & \underline{\text{x } 2} \\ 1080 & + \ 72 = 1152 \end{array}$$

We have the final answer 1152.

Example 2
Let us try another example. Multiply 26 by 23.

$$\begin{array}{c} 26 \\ \underline{\text{x } 23} \\ - - - - \end{array}$$

Step 1 - Break the multiplier.

$$\begin{array}{c} 26 \\ \underline{\text{x } 23} = (20 + 3) \\ - - - - \end{array}$$

So 23 is broken down as 20 + 3.

$$\begin{array}{c} 26 \\ \underline{\text{x } 20} + 3 \\ - - - - \end{array}$$

Step 2 - Multiply from left to right. Multiplying 26 x 20 = 520.

$$\begin{array}{c} 26 \\ \underline{\text{x } 20} + 3 \\ 520 \end{array}$$

Multiplying 26 x 3 = 78.

$$\begin{array}{cc} \mathbf{26} & \mathbf{26} \\ \underline{\mathbf{x\ 20}} & + \underline{\mathbf{x\ 3}} \\ \mathbf{520} & \mathbf{78} \end{array}$$

Step 3 - Add the individual answers together.

Adding 520 + 78 = 598.

$$\begin{array}{cc} \mathbf{26} & \mathbf{26} \\ \underline{\mathbf{x\ 20}} & \underline{\mathbf{x\ 3}} \\ \mathbf{520} & + \ \mathbf{78} = \mathbf{598} \end{array}$$

And we have the final answer 593.

Example 3

Let us try another example. Let us multiply 72 by 41.

$$\begin{array}{c} \mathbf{72} \\ \underline{\mathbf{x\ 41}} \\ \mathbf{-\ -\ -\ -} \end{array}$$

Take a second to apply the technique by yourself as fast as you can. Once you have the answer, you can check the steps below to see if you got your answer right.

Step 1 - Break the multiplier.

$$\begin{array}{c} \mathbf{72} \\ \underline{\mathbf{x\ 41}} = \mathbf{(40 + 1)} \\ \mathbf{-\ -\ -\ -} \end{array}$$

41 is broken down as 40 + 1.

$$\begin{array}{c} \mathbf{72} \\ \underline{\mathbf{x\ 40}} + \mathbf{1} \\ \mathbf{-\ -\ -\ -} \end{array}$$

Step 2 - Multiply from left to right.

Multiplying 72 x 40 gives 2880.

$$\begin{array}{r} 72 \\ \underline{\text{x 40}} + 1 \\ 2880 \end{array}$$

Multiplying 72 x 1 gives 72.

$$\begin{array}{r} 72 \\ \underline{\text{x 40}} \\ 2880 \end{array} + \begin{array}{r} 72 \\ \underline{\text{x 1}} \\ 72 \end{array}$$

Step 3 - Add the individual answers together.

Adding 2880 + 72 = 2952.

$$\begin{array}{r} 72 \\ \underline{\text{x 40}} \\ 2880 \end{array} + \begin{array}{r} 72 \\ \underline{\text{x 1}} \\ 72 \end{array} = 2952$$

If you got your answer wrong, don't worry. Just revisit the techniques and examples we covered in this chapter.

Students of the Ofpad Mental Math Course **ofpad.com/mathcourse** can post their questions in the discussion section. If you are not a student of the video course, you can email your questions, suggestions & feedback to **hq@ofpad.com**.

Exercises

Download the rich PDFs for these exercises from **ofpad.com/mathexercises**.

01) 43 × 11	07) 42 × 36	13) 59 × 98
02) 36 × 71	08) 60 × 22	14) 50 × 90
03) 89 × 14	09) 72 × 47	15) 99 × 13
04) 29 × 48	10) 19 × 91	16) 89 × 25
05) 32 × 24	11) 56 × 19	17) 13 × 77
06) 36 × 24	12) 13 × 98	18) 53 × 24

Answers

01) 473	07) 1,512	13) 5,782
02) 2,556	08) 1,320	14) 4,500
03) 1,246	09) 3,384	15) 1,287
04) 1,392	10) 1,729	16) 2,225
05) 768	11) 1,064	17) 1,001
06) 864	12) 1,274	18) 1,272

LR Multiplication – Two Digit Multiplier With Round Up

Rounding up and multiplying is also useful for two-digit multipliers when the numbers end in 7, 8 or 9.

Method

To do that:

Step 1 - Round up a number.

Step 2 - Multiply the rounded up value and the amount you rounded up from left to right.

Step 3 - Subtract the two numbers.

Example 1

Let us look at an example. Multiply 87 by 99.

$$\begin{array}{r} 87 \\ \underline{\times\,99} \\ ----\end{array}$$

Step 1 - Round up a number.

$$\begin{array}{r} 87 \\ \underline{\times\,99} \quad (100-1) \\ ----\end{array}$$

So we round up 99 to 100. We rounded up by 1.

$$\begin{array}{r} 87 \\ \underline{\times\,100} -1 \\ ----\end{array}$$

Step 2 - Multiply the rounded up value and the amount you rounded up from left to right.

So multiply 87 with 100 to get 8700.

$$\begin{array}{r} 87 \\ \underline{\times\,100} - 1 \\ 8700 \end{array}$$

Multiply 87 with 1 to get 87.

$$\begin{array}{rcr} 87 && 87 \\ \underline{\times\,100} & - & \underline{\times\,1} \\ 8700 && 87 \end{array}$$

Step 3 - Subtract the two numbers.

Subtract 8700 – 87 to get 8613.

$$\begin{array}{rcl} 87 && 87 \\ \underline{\times\,100} && \underline{\times\,1} \\ 8700 & - & 87 = 8613 \end{array}$$

If you have trouble with this subtraction step remember that the last digits (7 in 87 and 3 in 8613) add to give 10 and the other digits (8 in 87 and 1 in 8613) add to gives 9.

$$\begin{array}{r} 87 \\ \underline{\times\,99} \\ 8613 \end{array}$$

Your final answer is 8613.

Example 2
Let us look at another example.

Multiply 41 by 57.

$$\begin{array}{r} 41 \\ \underline{\times\,57} \\ - - - - \end{array}$$

Step 1 - Round up a number.

$$41$$
$$\underline{\times 57} \quad (60 - 3)$$
$$- - - -$$

So we round up 57 to 60. We rounded up by 3.

$$41$$
$$\underline{\times 60} - 3$$
$$- - - -$$

Step 2 - Multiply the rounded up value and the amount you rounded up from left to right.

So multiply 41 with 60 to give 2460.

$$41$$
$$\underline{\times 60} - 3$$
$$2460$$

Then multiply 41 with 3 to give 123.

$$41 \qquad\qquad 41$$
$$\underline{\times 60} \quad - \quad \underline{\times 3}$$
$$2460 \qquad\quad 123$$

Step 3 - Subtract the two numbers.

Subtract 2460 − 123 to get 2337.

$$41 \qquad\qquad 41$$
$$\underline{\times 60} \qquad\quad \underline{\times 3}$$
$$2460 \quad - \quad 123 \; = 2337$$

Your final answer is 2337.

$$
\begin{array}{r}
41 \\
\times\,57 \\
\hline
2337
\end{array}
$$

Example 3

Let us look at another example.

Multiply 67 by 38.

$$
\begin{array}{r}
67 \\
\times 38 \\
\hline
\text{\textemdash\ \textemdash\ \textemdash\ \textemdash}
\end{array}
$$

Take a second to apply the technique by yourself as fast as you can. Once you have the answer, you can check the steps below to see if you got your answer right.

Step 1 - Round up a number.

$$
\begin{array}{r}
67 \\
\times 38 \quad (40 - 2) \\
\hline
\text{\textemdash\ \textemdash\ \textemdash\ \textemdash}
\end{array}
$$

So we round up 38 to 40. We rounded up by 2.

$$
\begin{array}{r}
67 \\
\times 40 - 2 \\
\hline
\text{\textemdash\ \textemdash\ \textemdash\ \textemdash}
\end{array}
$$

Step 2 - Multiply the rounded up value and the amount you rounded up from left to right.

Multiplying 67 by 40 gives 2680.

$$
\begin{array}{r}
67 \\
\times 40 - 2 \\
\hline
2680
\end{array}
$$

Next multiplying 67 by 2 gives 134.

$$
\begin{array}{rr}
\mathbf{67} & \mathbf{67} \\
\underline{\mathbf{x\,40}} & \underline{\mathbf{x\,2}} \\
\mathbf{2680} & \mathbf{134}
\end{array}
$$

Step 3 - Subtract the two numbers.

Subtracting 2680 − 134 gives 2546.

$$
\begin{array}{rr}
\mathbf{67} & \mathbf{67} \\
\underline{\mathbf{x\,40}} & \underline{\mathbf{x\,2}} \\
\mathbf{2680} & \mathbf{-\quad 134 \;=\; 2546}
\end{array}
$$

Now you have 2546 which is your final answer.

$$
\begin{array}{r}
\mathbf{67} \\
\underline{\mathbf{x\,38}} \\
\mathbf{2546}
\end{array}
$$

If you got your answer wrong, don't worry. Just revisit the techniques and examples we covered in this chapter.

Students of the Ofpad Mental Math Course **ofpad.com/mathcourse** can post their questions in the discussion section. If you are not a student of the video course, you can email your questions, suggestions & feedback to **hq@ofpad.com**.

Exercises

Download the rich PDFs for these exercises from **ofpad.com/mathexercises**.

01) 67 x 76	07) 38 x 38	13) 58 x 38
02) 38 x 82	08) 58 x 83	14) 18 x 96
03) 29 x 29	09) 18 x 90	15) 37 x 65
04) 67 x 61	10) 69 x 36	16) 49 x 22
05) 19 x 91	11) 67 x 42	17) 79 x 17
06) 49 x 54	12) 79 x 84	18) 49 x 34

Answers

01) 5,092	09) 1,620	17) 1,343
02) 3,116	10) 2,484	18) 1,666
03) 841	11) 2,814	
04) 4,087	12) 6,636	
05) 1,729	13) 2,204	
06) 2,646	14) 1,728	
07) 1,444	15) 2,405	
08) 4,814	16) 1,078	

LR Multiplication After Factoring

Like rounding up, another technique to use before you apply the LR method is to factor the number before multiplying it. Factoring a number means breaking it down into one digit numbers, which when later multiplied together will give the original number.

For example, 48 can be factored as 6 x 8 or 12 x 4. However, we will prefer to use single digit factors as the intention of factoring the numbers is to simplify the multiplication.

Method

To apply LR method with factoring:

Step 1 - Factor one of the numbers.

Step 2 - Multiply the number with the first factor (left to right).

Step 3 - Multiply the product with the second factor (left to right).

Example 1

Multiply 45 by 22.

$$\begin{array}{r} 45 \\ \underline{\times\ 22} \\ - - - \end{array}$$

Step 1 - Factor one of the numbers.

Let us factor 22. So 22 can be factored as 11 x 2.

$$\begin{array}{r} 45 \\ \underline{\times\ 11} \quad \times 2 \\ - - - \end{array}$$

Step 2 - Multiply the number with the first factor (left to right).

So multiply 45 by 11. You get 495. Since we are multiplying by 11, you can apply the rule to multiply by 11 instead of doing an LR multiplication.

$$\begin{array}{r} 45 \\ \underline{\times\ 11} \\ 495 \end{array} \qquad \textbf{x2}$$

Step 3 - Multiply the product with the second factor (left to right).

So now take the 495 and multiply it by 2. Multiply from left to right. So 495 multiplied by 2 gives 990.

$$\begin{array}{r} 495 \\ \underline{\times\ 2} \\ 990 \end{array}$$

And 990 is your final answer.

You could have factored 45 as 9 x 5 and multiplied it by 22 instead of factoring 22. Also, you could have used 2 as the first factor and 11 as the second factor. The choice is always up to you, but the easier and smaller the numbers, the faster you will calculate.

Example 2
Let us look at another example.

Multiply 21 by 63.

$$\begin{array}{r} 21 \\ \underline{\times\ 63} \\ - - - \end{array}$$

Step 1 - Factor one of the numbers.

Let us factor 63. So 63 can be factored as 7 x 9.

$$
\begin{array}{r}
21 \\
\underline{\times\ 7} \quad \times 9 \\
-\,-\,-
\end{array}
$$

Step 2 - Multiply the number with the first factor (left to right).

So multiply 21 by 7 left to right. You get 147.

$$
\begin{array}{r}
21 \\
\underline{\times\ 7} \quad \times 9 \\
147
\end{array}
$$

Step 3 - Multiply the product with the second factor (left to right).

So now take the 147 and multiply it by 9. Multiply from left to right.

So 147 multiplied by 9 gives 1323.

$$
\begin{array}{r}
147 \\
\underline{\times\ 9} \\
1323
\end{array}
$$

Your final answer is 1323.

You could have factored 21 into 7 x 3 and solved this problem, and some would have found that easier.

Example 3

Let us look at another example.

Multiply 42 by 36.

$$42$$
$$\underline{\times\ 36}$$
$$-\ -\ -$$

Take a second to apply the technique by yourself. Once you have the answer, you can check the steps below to see if you got your answer right.

Step 1 - Factor one of the numbers.

Let us factor 36. So 36 can be factored as 6 x 6.

$$42$$
$$\underline{\times\ 6}\qquad \times 6$$
$$-\ -\ -$$

Step 2 - Multiply the number with the first factor (left to right).

So multiply 42 by 6 left to right. You get 252.

$$42$$
$$\underline{\times\ 6}\qquad \times 6$$
$$252$$

Step 3 - Multiply the product with the second factor (left to right).

So now take the 252 and multiply it by 6. Multiply from left to right. So 252 multiplied by 6 gives 1512.

$$252$$
$$\underline{\times\ \ 6}$$
$$1512$$

And 1512 is your final answer.

You could have factored 42 into 7 x 6 instead of factoring 36 into 6 x 6 and you would have got the same answer. If you got your answer wrong, don't worry. Just revisit the techniques and examples we covered in this chapter.

You can post your questions in the discussion section of **ofpad.com/mathcourse** or email me at **hq@ofpad.com**. If you have enjoyed the book so far, do leave a review on Amazon by visiting **ofpad.com/mathbook**. Practice before proceeding.

Exercises
Download the rich PDFs for these exercises from **ofpad.com/mathexercises**.

01) 38 x 77	07) 28 x 36	13) 79 x 72
02) 37 x 63	08) 68 x 33	14) 38 x 56
03) 37 x 54	09) 58 x 16	15) 87 x 24
04) 89 x 54	10) 27 x 81	16) 17 x 42
05) 49 x 99	11) 77 x 24	17) 67 x 64
06) 89 x 56	12) 77 x 49	18) 17 x 18

Answers
01) 2,926
02) 2,331
03) 1,998
04) 4,806
05) 4,851
06) 4,984
07) 1,008
08) 2,244
09) 928
10) 2,187
11) 1,848
12) 3,773
13) 5,688
14) 2,128
15) 2,088
16) 714
17) 4,288
18) 306

Chapter 9 - Math Anxiety

When some of us try to do the math, we have butterflies in our stomach. This sometimes also gives us sweaty palms and makes it hard to concentrate.

This phenomenon is called math anxiety, and most of us suffer from it in varying degrees. You might think you are anxious about math because you are bad at it. But it's often the other way around. You are doing poorly in math because you are anxious about it.

Math anxiety decreases your working memory. Worrying about being able to solve a math problem, eats up working memory leaving less of it available to tackle math itself.

So we can struggle with even the basic math skills that we have otherwise mastered. Relaxing when doing math is therefore essential.

The secrets of doing mental math in this book are going to change the way you do math forever. This is going to make doing math in your head as easy as reading a comic.

But your new math superpower is not going to make you a superhero unless you relax. If you ever feel math anxiety, take a few deep breaths, and you will be able to do the math faster.

Use the knowledge about how your working memory works to create a mindset where you enjoy doing the math. See every math problem as a challenge that will increase your intelligence.

See yourself not only doing math faster but also enjoying yourself while doing it. Do this, and you will increase your math speed even further.

Chapter 10 - Squaring

In this chapter, we will look at how to square numbers fast. Squaring a number means multiplying a number by itself (example 5 x 5 or 43 x 43). It is usually written as 5^2 or $5\wedge 2$.

Squaring Numbers Ending With 5

Let us first look at squaring numbers which end with 5 first. We will look at squaring any number next, and it uses the same principle as squaring number ending with 5.

Method

To square a number ending with 5:

Step 1 - Multiply the first digit with the next higher digit to get the first few digits of the answer.

Step 2 - Attach 25 to the first few digits of your answer to get the full answer.

Example 1

Let us look at an example. Square 65.

$$
\begin{array}{r}
65 \\
\underline{\times\ 65} \\
\_\ \_\ \_\ \_
\end{array}
$$

Step 1 - Multiply the first digit with the next higher digit to get the first few digits of the answer.

So the first digit 6 is multiplied by the next higher digit 7 to give 42.

$$
\begin{array}{r}
65 \\
\underline{\times\ 65} \\
42\ \_\ \_
\end{array}
$$

OLD NAVY 4216
401 NE NORTHGATE WAY
Seattle WA 98125
(206) 417 4694

12/24/2017   3:08:19 PM
Trans: 9713   Store: 04216
Reg: 006
Cashier: 2720002

Complete a survey within 5 days from
today and receive 10% off your next
in store and online purchase at Old Navy
plus free shipping when you use your
discount to shop online
Take the survey at
www.feedback@oldnavy.com

Completa una encuesta dentro de 5 días a
partir de hoy y recibe un 10% de
descuento en tu próxima compra en tienda
y en oldnavy.com. Además, recibe envío
gratis cuando utilizas este descuento en
línea.
Ve a www.feedback@oldnavy.com

At the end of the survey you will be
given a code to record here
Offer applies to merchandise only. No
price adjustments on prior purchases
Cannot be combined with other offers
Offer expires 3 months from today

OLD NAVY 4216
401 NE NORTHGATE WAY
Seattle, WA 98125
(206) 417-4694

12/26/2017   3:08:19 PM
Trans.: 9713   Store: 04216
Reg.: 006
Cashier: 2706002

Complete a survey within 5 days from
today and receive 10% off your next
in store and online purchase at Old Navy
plus free shipping when you use your
discount to shop online.
Take the survey at
www.feedback4oldnavy.com

Completa una encuesta dentro de 5 dias a
partir de hoy y recibe un 10% de
descuento en tu proxima compra en tienda
y en oldnavy.com. Ademas, recibe envoi
gratis cuando utilizas este descuento en
linea.
Ve a www.feedback4oldnavy.com

At the end of the survey you will be
given a code to record here:
Offer applies to merchandise only. No
price adjustments on prior purchases.
Cannot be combined with other offers.
Offer expires 3 months from today.

Step 2 - Attach 25 to the first few digits of your answer to get the full answer.

Attaching 25 to 42 we get 4225.

$$
\begin{array}{r}
65 \\
\times\ 65 \\
\hline
4225
\end{array}
$$

Your final answer is 4225.

Example 2
Square 45.

$$
\begin{array}{r}
45 \\
\times\ 45 \\
\hline
\_\ \_\ \_\ \_
\end{array}
$$

Step 1 - Multiply the first digit with the next higher digit to get the first few digits of the answer.

So the first digit 4 multiplied by the next higher digit 5 gives 20.

$$
\begin{array}{r}
45 \\
\times\ 45 \\
\hline
20\ \_\ \_
\end{array}
$$

Step 2 - Attach 25 to the first few digits of your answer to get the full answer.

Attaching 25 to 20 we get 2025.

$$
\begin{array}{r}
45 \\
\times\ 45 \\
\hline
2025
\end{array}
$$

Attaching 25 you get the final answer as 2025.

Example 3

Square the number 85.

$$85$$
$$\underline{\times\ 85}$$

$$-\ -\ -\ -$$

Take a second to apply the technique by yourself as fast as you can. Once you have the answer, you can check the steps below to see if you got your answer right.

Step 1 - Multiply the first digit with the next higher digit to get the first few digits of the answer.

So the first digit 8 is multiplied by the next higher digit 9 to give 72.

$$85$$
$$\underline{\times\ 85}$$
$$72\ \_\ \_$$

Step 2 - Attach 25 to this number to get the full answer.

$$85$$
$$\underline{\times\ 85}$$
$$7225$$

So we have the final answer as 7225.

If you got your answer wrong, don't worry. Just revisit the techniques and examples we covered in this chapter.

Students of the Ofpad Mental Math Course **ofpad.com/mathcourse** can post their questions in the discussion section. If you are not a student of the video course, you can email your questions, suggestions & feedback to **hq@ofpad.com**.

Squaring Any Number

Numbers that you want to square does not always end with 5. So we will look at the general method of squaring numbers.

Step 1 - Round the number up or down to the nearest multiple of 10.

Step 2 - If you rounded up, subtract the number by the amount by which you rounded up. If you rounded down, add instead of subtract.

Step 3 - Multiply the numbers from the previous two steps.

Step 4 - Square the amount you rounded up/down and add it to the number from the previous step.

The squaring of numbers ending with 5 is just a simplification of this method.

Example 1

Let us look at an example to understand the method further.

Square the number 23.

$$\begin{array}{r} 23 \\ \times\,23 \\ \hline --- \end{array}$$

Step 1 - Round the number up or down to the nearest multiple of 10. So round down 23 to 20.

$$\begin{array}{r} 23 \\ \underline{\times\,23} - 3 = 20 \\ --- \end{array}$$

Step 2 - If you rounded up, subtract the number by the amount by which you rounded up. If you rounded down, add instead of subtract.

Since we rounded down, we add the amount we rounded down, which is 3 to get 26.

$$23 + 3 = 26$$
$$\underline{x\ 23} - 3 = 20$$

$$- - - -$$

Step 3 - Multiply the numbers from the previous two steps. Remember to multiply from left to right.

Multiply 26 with 20 to get 520.

$$23 + 3 = 26$$
$$\underline{x\ 23} - 3 = \underline{x\ 20}$$
$$520$$

Step 4 - Square the amount you rounded up/down and add it to the number from the previous step.

We rounded down 23 by 3. Squaring 3, we get 9.

$$23 +\ \ 3 = \ \ 26$$
$$\underline{x\ 23} - \underline{x\ 3} = \underline{x20}$$
$$9 \quad 520$$

Adding 9 to 520 we get 529.

$$23 +\ \ 3 = 26$$
$$\underline{x23} - \underline{x3} = \underline{x20}$$
$$9 + 520 = 529$$

And you have the final answer of the square of 23 as 529.

$$\begin{array}{r} 23 \\ \times\ 23 \\ \hline 529 \end{array}$$

Example 2

Square the number 48.

$$\begin{array}{r} 48 \\ \times\ 48 \\ \hline - - - \end{array}$$

Step 1 - Round the number up or down to the nearest multiple of 10.

So round up 48 to 50.

$$\begin{array}{r} 48 \\ \underline{\times 48} + 2 = 50 \\ - - - \end{array}$$

Step 2 - If you rounded up, subtract the number by the amount by which you rounded up. If you rounded down, add instead of subtract.

Since we rounded up, we subtract the amount we rounded up, which is 2 to get 46.

$$\begin{array}{r} 48\ -\ 2\ =\ 46 \\ \underline{\times\ 48}\ +\ 2\ =\ 50 \\ - - - \end{array}$$

Step 3 - Multiply the numbers from the previous two steps.

Multiply 46 with 50 to get 2300. Remember to multiply from left to right.

$$48 \quad - \quad 2 = \quad 46$$
$$\underline{x\,48} \quad + \quad 2 = \quad \underline{x50}$$
$$2300$$

Step 4 - Square the amount you rounded up/down and add it to the number from the previous step.

We rounded up 48 by 2. Squaring 2, we get 4.

$$48 \quad - \quad 2 = \quad 46$$
$$\underline{x\,48} + \underline{x2} = \quad \underline{x\,50}$$
$$4 \qquad 2300$$

Adding it to 2300, we get 2304.

$$48 \quad - \quad 2 = \quad 46$$
$$\underline{x\,48} + \underline{x\,2} = \underline{x\,50}$$
$$4 + 2300 = 2304$$

So the square of 48 is 2304 which is your final answer.

$$48$$
$$\underline{x\,48}$$
$$2304$$

Example 3
Square the number 64.

$$64$$
$$\underline{x\,64}$$

— — —

Take a second to apply the technique by yourself as fast as you can. Once you have the answer, you can check the steps below to see if you got your answer right.

Step 1 - Round the number up or down to the nearest multiple of 10.

So round down 64 to 60.

$$\begin{array}{r} 64 \\ \underline{\text{x } 64} - 4 = 60 \\ \text{---} \end{array}$$

Step 2 - If you rounded up, subtract the number by the amount by which you rounded up. If you rounded down, add instead of subtract.

Since we rounded down, we add the amount we rounded down, which is 4 to get 68.

$$\begin{array}{r} 64 \; + \; 4 \; = \; 68 \\ \underline{\text{x } 64} \; - \; 4 \; = \; 60 \\ \text{---} \end{array}$$

Step 3 - Multiply the numbers from the previous two steps. Remember to multiply from left to right.

Multiply 68 with 60 to get 4080.

$$\begin{array}{r} 64 \; + \; 4 \; = \;\;\; 68 \\ \underline{\text{x}64} \; - \; 4 \; = \; \underline{\text{x}60} \\ 4080 \end{array}$$

Step 4 - Square the amount you rounded up/down and add it to the number from the previous step.

We rounded down 64 by 4. Squaring 4, we get 16.

$$\begin{array}{r} 64 \; + \; 4 \; = \;\;\; 68 \\ \underline{\text{x}64} \; - \; \underline{\text{x}4} \; = \underline{\text{x}60} \\ 16 \;\;\; 4080 \end{array}$$

Adding it to 4080, we get 4096.

$$64 \; + \; 4 \; = \; 68$$
$$\underline{x \, 64} \; - \; \underline{x4} \; = \underline{x \, 60}$$
$$16 + 4080 = 4096$$

You have now got the square of 64 as 4096.

$$64$$
$$\underline{x \, 64}$$
$$4096$$

If you got your answer wrong, don't worry. Just revisit the techniques and examples we covered in this chapter.

Students of the Ofpad Mental Math Course **ofpad.com/mathcourse** can post their questions in the discussion section. If you are not a student of the video course, you can email your questions, suggestions & feedback to **hq@ofpad.com**.

Read this chapter again if necessary.

Then go to the practice section and complete the exercises.

You might have understood the technique, but it will take practice before the technique becomes second nature to you.

Only practice will make using these techniques effortless and easy.

If you have enjoyed the book so far, do leave a review on Amazon by visiting **ofpad.com/mathbook**, so that others might also benefit from reading this book.

Squaring Exercises

Download the rich PDFs for these exercises from **ofpad.com/mathexercises**.

01) 35 x 35	07) 56 x 56	13) 85 x 85
02) 73 x 73	08) 71 x 71	14) 15 x 15
03) 46 x 46	09) 39 x 39	15) 53 x 53
04) 97 x 97	10) 63 x 63	16) 330 x 330
05) 24 x 24	11) 83 x 83	17) 745 x 745
06) 83 x 83	12) 28 x 28	18) 515 x 515

Squaring Answers

01) 1,225	07) 3,136	13) 7,225
02) 5,329	08) 5,041	14) 225
03) 2,116	09) 1,521	15) 2,809
04) 9,409	10) 3,969	16) 108,900
05) 576	11) 6,889	17) 555,025
06) 6,889	12) 784	18) 265,225

Chapter 11 - The Bridge Method

In this chapter, we will look at how to do fast multiplication using the direct method. You will be able to multiply two and three digit numbers together using this method.

To multiply using the bridge method:

Step 1 - Multiply the first number of the multiplicand with the first number of the multiplier and put the result as the left-hand number of the answer.

Step 2 - Multiply the outside pairs and multiply the inside pairs.

Step 3 - Add the products of the outside and inside pairs together to get the next figure of the answer.

Step 4 - Multiply the last number of the multiplicand with the second number of the multiplier and put the result as the right-hand number of the answer.

Tip - Use your forefinger and middle finger to pace through the multiplication problem. Place them on the screen/paper, so you keep track of which numbers you are calculating.

Let us look at an example.

Example 1
Multiply 32 by 13.

Use your forefinger and middle finger to pace through the multiplication problem. Place them on the screen/paper, so you keep track of which numbers you are calculating.

$$3\ 2\ x\ 1\ 3$$

Step 1 - Multiply the first number of the multiplicand with the first number of the multiplier and put the result as the left-hand number of the answer.

Multiply 3 with 1 to give 3.

$$3\,2 \times 1\,3 = 3\,\_\,\_$$

Step 2 - Multiply the outside pairs and multiply the inside pairs.

So multiplying the outside pairs 3 and 3, we get 9.

Multiplying the inside pairs 2 and 1, we get 2.

$$
\begin{array}{c}
3 \quad 2 \times 1 \quad 3 = 3\,\_\,\_ \\
9 \\
2
\end{array}
$$

Step 3 - Add the products of the outside and inside pairs together to get the next figure of the answer

Adding 9 and 2, we get 11.

$$
\begin{array}{c}
3 \quad 2 \times 1 \quad 3 = 3\,\_\,\_ \\
9 \\
\underline{+2} \\
11
\end{array}
$$

Carry over the one so that 3 becomes 4.

$$3 \quad 2 \times 1 \quad 3 = 4\,1\,\_$$

Step 4 - Multiply the second number of the multiplicand with the second number of the multiplier and put the result as the right-hand number of the answer.

So multiplying 2 and 3 we get 6 which is the last number of the answer.

3 2 x 1 3 = 4 1 6

The final answer is 416.

Example 2
Multiply 323 with 13.

Use your forefinger and middle finger to pace through the multiplication problem. Place them on the screen/paper, so you keep track of which numbers you are calculating.

3 2 3 x 1 3

Step 1 - Multiply the first number of the multiplicand with the first number of the multiplier and put the result as the left-hand number of the answer.

Multiply 3 with 1 to give 3.

3 2 3 x 1 3 = 3 _ _ _

Step 2 - Multiply the outside pairs and multiply the inside pairs.

So multiplying the outside pairs 3 and 3, we get 9 and

Multiplying the inside pairs 2 and 1, we get 2.

$$3\ 2\ 3\ x\ 1\ 3 = 3\ \_\ \_\ \_$$
$$9$$
$$2$$

Step 3 - Adding the products of the outside and inside pairs together we get the next figure of the answer

Adding 2 and 9 we get 11.

$$3\ 2\ 3\ x\ 1\ 3 = 3\ \_\ \_\ \_$$
$$9$$
$$\underline{+2}$$
$$11$$

Carry over the one so that 3 becomes 4.

$$3\ 2\ 3\ x\ 1\ 3 = 4\ 1\ \_\ \_$$

Repeat the steps for the next set of digits.

Step 2 - Multiply the outside pairs and multiply the inside pairs.

So multiplying the outside pairs 2 and 3, we get 6 and

Multiplying the inside pairs 3 and 1, we get 3.

$$3\ 2\ 3\ x\ 1\ 3 = 4\ 1\ \_\ \_$$
$$6$$
$$3$$

Step 3 - Add the products of the outside and inside pairs together to get the next figure of the answer

$$323 \times 13 = 41\_\_$$
$$6$$
$$\underline{+3}$$
$$9$$

Adding 6 and 3, we get 9.

$$323 \times 13 = 419\_$$

Step 4 - Multiply the last number of the multiplicand with the second number of the multiplier and put the result as the right-hand number of the answer.

So multiplying 3 and 3 we get 9 which is the last number of the answer.

$$323 \times 13 = 4199$$

The answer is 4199.

Example 3
Multiply 323 by 132

$$323 \times 132$$

Take a second to apply the technique by yourself as fast as you can. Once you have the answer, you can check the steps below to see if you got your answer right. Remember to use your forefinger and middle finger to pace through the multiplication problem

Step 1 - Multiply the first number of the multiplicand with the first number of the multiplier and put the result as the left-hand number of the answer.

Multiply 3 with 1 to give 3.

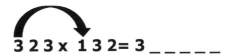

3 2 3 x 1 3 2 = 3 _ _ _ _ _

Step 2 - Multiply the outside pairs and multiply the inside pairs.

So multiplying the outside pairs 3 and 3 we get 9 and

Multiplying the inside pairs 2 and 1 we get 2.

3 2 3 x 1 3 2 = 3 _ _ _ _
                          9
                          2

Step 3 - Add the products of the outside and inside pairs together to get the next figure of the answer.

Adding 9 and 2 we get 11.

3 2 3 x 1 3 2 = 3 _ _ _ _
                          9
                        +2
                         11

Carry over the 1 so the 3 becomes 4.

3 2 3 x 1 3 2 = 4 1 _ _ _

Repeat the steps for the next set of digits

Step 2 - Multiply the outside pairs and multiply the inside pairs.

a) Multiplying the outside pairs 3 and 2 we get 6
b) Multiplying the inside pairs 2 and 3 we get 6
c) Multiplying the last set of inner pairs 3 and 1 we get 3.

$$3\ 2\ 3\ x\ 1\ 3\ 2 = 4\ 1\ \_\ \_\ \_$$
$$6$$
$$6$$
$$\underline{+3}$$

Step 3 - Add the products of the outside and inside pairs together to get the next figure of the answer

Adding 6 with 6 and then adding it with 3 we get 15.

$$3\ 2\ 3\ x\ 1\ 3\ 2 = 4\ 1\ \_\ \_\ \_$$
$$6$$
$$6$$
$$\underline{+3}$$
$$15$$

Carry over the one, so the 1 becomes 2.

$$3\ 2\ 3\ x\ 1\ 3\ 2 = 4\ 2\ 5\ \_\ \_$$

Repeat the steps for the next set of digits

Step 2 - Multiply the outside pairs and multiply the inside pairs.

So multiplying the outside pairs 2 and 2 we get 4,

Multiplying the inside pairs 3 and 3 we get 9.

$$3\ 2\ 3\ x\ 1\ 3\ 2 = 4\ 2\ 5\ \_\ \_$$
$$4$$
$$\underline{+9}$$

Step 3 - Add the products of the outside and inside pairs together to get the next figure of the answer

Adding 4 with 9 we get 13.

$$3\,2\,3 \times 1\,3\,2 = 4\,2\,5\_\_$$
$$4$$
$$\underline{+9}$$
$$13$$

Carry over the one, so the 5 becomes 6.

$$3\,2\,3 \times 1\,3\,2 = 4\,2\,6\,3\_$$

Step 4 - Multiply the last number of the multiplicand with the last number of the multiplier and put the result as the right-hand number of the answer.

So multiplying 3 and 2 we get 6 which is the last number of the answer.

$$3\,2\,3 \times 1\,3\,2 = 4\,2\,6\,3\,6$$

The answer is 42,636.

If you got your answer wrong, don't worry. Just revisit the techniques and examples we covered in this chapter.

Students of the Ofpad Mental Math Course **ofpad.com/mathcourse** can post their questions in the discussion section. If you are not a student of the video course, you can email your questions, suggestions & feedback to **hq@ofpad.com**.

Read this chapter again if necessary. Then go to the practice section and complete the exercises.

You might have understood the technique, but it will take practice before the technique becomes second nature to you.

Only practice will make using these techniques effortless and easy.

If you have enjoyed the book so far, do leave a review on Amazon by visiting **ofpad.com/mathbook**, so that others might also benefit from reading this book.

Exercises
Download the rich PDFs for these exercises from **ofpad.com/mathexercises**.

01) 38 x 74	07) 917 x 13	13) 368 x 349
02) 63 x 63	08) 140 x 66	14) 106 x 783
03) 14 x 22	09) 389 x 60	15) 312 x 642
04) 12 x 57	10) 852 x 94	16) 430 x 152
05) 26 x 95	11) 459 x 92	17) 970 x 639
06) 25 x 87	12) 687 x 89	18) 553 x 963

Answers

01) 2,812	07) 11,921	13) 128,432
02) 3,969	08) 9,240	14) 82,998
03) 308	09) 23,340	15) 200,304
04) 684	10) 80,088	16) 65,360
05) 2,470	11) 42,228	17) 619,830
06) 2,175	12) 61,143	18) 532,539

Chapter 12 - Vitruvian Man Method

It is called the Vitruvian Man method because the visualisation methods resemble that of a Vitruvian man. The steps in the method are same as the bridge method. The only thing which is different is the mental visualisation.

Use your forefinger and middle finger to pace through the multiplication problem.

Example 1
Multiply 32 by 13.

$$3\ 2$$

$$x\ 1\ 3$$

Let us apply the new method of visualisation.

Use your forefinger and middle finger to pace through the multiplication problem.

Step 1 - Multiply the first number of the multiplicand with the first number of the multiplier and put the result as the left-hand number of the answer.

Multiply 3 with 1 to give 3.

Step 2 - Multiply the outside pairs and multiply the inside pairs.

So multiplying the outside pairs 3 and 3 we get 9.

Multiplying the inside pairs 2 and 1 we get 2.

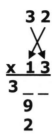

```
    3 2
   x 1 3
    3 _ _
      9
      2
```

Step 3 - Add the products of the outside and inside pairs together to get the next figure of the answer

Adding 2 and 9, we get 11.

```
    3 2
   x 1 3
    3 _ _
      9
     +2
     11
```

Carry over the one so that 3 becomes 4.

```
    3 2
   x 1 3
    41 _
```

Step 4 - Multiply the second number of the multiplicand with the second number of the multiplier and put the result as the right-hand number of the answer.

So multiplying 2 and 3 we get 6 which is the last number of the answer.

The answer is 416.

Example 2
Multiply 323 with 13.

$$323$$

$$x \quad 13$$

$$\overline{- - - -}$$

Apply the new method of visualisation.

Use your forefinger and middle finger to pace through the multiplication problem.

Step 1 - Multiply the first number of the multiplicand with the first number of the multiplier and put the result as the left-hand number of the answer.

Multiply 3 with 1 to give 3.

Step 2 - Multiply the outside pairs and multiply the inside pairs.

So multiplying the outside pairs 3 and 3 we get 9.

Multiplying the inside pairs 2 and 1 we get 2.

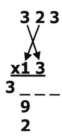

Step 3 - Add the products of the outside and inside pairs together to get the next figure of the answer.

Adding 2 and 9, we get 11.

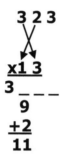

Carry over the one so that 3 becomes 4.

```
    3 2 3
    x1 3
  4 1 _ _
```

Repeat the steps for the next set of digits.

Step 2 - Multiply the outside pairs and multiply the inside pairs.

So multiplying the outside pairs 2 and 3, we get 6.

Multiplying the inside pairs 3 and 1, we get 3.

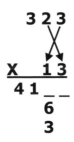

Step 3 - Add the products of the outside and inside pairs together to get the next figure of the answer

Adding 6 and 3, we get 9.

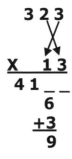

So 9 becomes the next digit of the answer.

Step 4 - Multiply the last number of the multiplicand with the second number of the multiplier and put the result as the right-hand number of the answer.

So multiplying 3 and 3 we get 9 which is the last number of the answer.

$$3\ 2\ 3$$

$$\text{X}\quad 1\ 3$$

$$4\ 1\ 9\ 9$$

The answer is 4199.

Example 3
Multiply 323 by 132

Apply the new method of visualisation.

3 2 3

x 1 3 2

_ _ _ _ _

Take a second to apply the technique by yourself as fast as you can. Once you have the answer, you can check the steps below to see if you got your answer right. Remember to use your forefinger and middle finger to pace through the multiplication problem

Step 1 - Multiply the first number of the multiplicand with the first number of the multiplier and put the result as the left-hand number of the answer.

Multiply 3 with 1 to give 3.

Step 2 - Multiply the outside pairs and multiply the inside pairs.

So multiplying the outside pairs 3 and 3 we get 9.

Multiplying the inside pairs 2 and 1 we get 2.

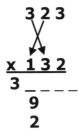

```
    3 2 3
       X
x   1 3 2
3 _ _ _ _
      9
      2
```

Step 3 - Add the products of the outside and inside pairs together to get the next figure of the answer.

Adding 9 and 2, we get 11.

```
    3 2 3
       X
x   1 3 2
3 _ _ _ _
      9
     +2
     11
```

Carry over the 1 so the 3 becomes 4.

```
    3 2 3
       X
x   1 3 2
4 1 _ _ _
```

Repeat the steps for the next set of digits

Step 2 - Multiply the outside pairs and multiply the inside pairs.

   a) So multiplying the outside pairs 3 and 2 we get 6.
   b) Multiplying the inside pairs 2 and 3 we get 6.
   c) Multiply the last set of inner pairs 3 and 1 to give 3.

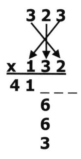

Step 3 - Add the products of the outside and inside pairs together to get the next figure of the answer

Adding 6 with 6 and then adding it with 3 we get 15.

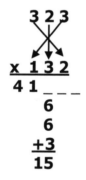

Carry over the one, so the 1 becomes 2.

```
    3 2 3
     X
 x 1 3 2
 4 2 5 _ _
```

Repeat the steps for the next set of digits

Step 2 - Multiply the outside pairs and multiply the inside pairs.

$$
\begin{array}{r}
3\ 2\ 3 \\
\times\ 1\ 3\ 2 \\
\hline
4\ 2\ 5\ \_\ \_
\end{array}
$$

So multiplying the outside pairs 2 and 2 we get 4.

Multiplying the inside pairs 3 and 3 we get 9.

$$
\begin{array}{r}
3\ 2\ 3 \\
\times\ 1\ 3\ 2 \\
\hline
4\ 2\ 5\ \_\ \_ \\
4 \\
9
\end{array}
$$

Step 3 - Add the products of the outside and inside pairs together to get the next figure of the answer

Adding 4 with 9 we get 13.

$$
\begin{array}{r}
3\ 2\ 3 \\
\times\ 1\ 3\ 2 \\
\hline
4\ 2\ 5\ \_\ \_ \\
4 \\
+\ 9 \\
\hline
13
\end{array}
$$

Carry over the one, so the 5 becomes 6.

Step 4 - Multiply the last number of the multiplicand with the last number of the multiplier and put the result as the right-hand number of the answer.

So multiplying 3 and 2 we get 6 which is the last number of the answer.

$$\begin{array}{r} 3\,2\,3 \\ \times\,1\,3\,2 \\ \hline 4\,2\,6\,3\,6 \end{array}$$

The answer is 42,636.

If you got your answer wrong, don't worry. Just revisit the techniques and examples we covered in this chapter.

Students of the Ofpad Mental Math Course **ofpad.com/mathcourse** can post their questions in the discussion section. If you are not a student of the video course, you can email your questions, suggestions & feedback to **hq@ofpad.com**.

Read this chapter again if necessary. Then go to the practice section and complete the exercises. You might have understood the technique, but it will take practice before the technique becomes second nature to you.

If you have enjoyed the book so far, do leave a review on Amazon by visiting **ofpad.com/mathbook**, so that others might also benefit from reading this book.

Exercises
Download the rich PDFs for these exercises from **ofpad.com/mathexercises**.

01) 97 x 79	07) 366 x 68	13) 252 x 881
02) 22 x 21	08) 139 x 25	14) 582 x 473
03) 41 x 32	09) 458 x 78	15) 406 x 926
04) 421 x 63	10) 306 x 56	16) 865 x 115
05) 692 x 38	11) 874 x 42	17) 396 x 487
06) 897 x 40	12) 422 x 80	18) 699 x 244

Answers

01) 7,663
02) 462
03) 1,312
04) 26,523
05) 26,296
06) 35,880
07) 24,888

08) 3,475
09) 35,724
10) 17,136
11) 36,708
12) 33,760
13) 222,012
14) 275,286

15) 375,956
16) 99,476
17) 192,852
18) 170,556

Chapter 13 - UT Method

This method is called UT method where UT is short for Units and Tens. This is another method of multiplication and you will be able to do large multiplication problems fast.

UT Method Concepts

There are a few concepts and definition you will have to understand before we apply the UT method:

1) A digit is a one figure number (e.g. 4, 2, 0).
2) Multiplying a digit by a digit will give you a one figure or two figure number but never longer (9 x 9 = 81).
3) Sometimes a digit multiplied by a digit will give a one figure number. In such cases for the UT method, it is treated as a 2 digit number by adding a 0 in front of it (e.g. 2 x 3 = 06).
4) In a 2 digit number, the left-hand digit is the tens digit (T) and the right-hand digit is the units digit (U) (e.g. In 23, 2 is T and 3 is U).
5) We will use either U or T of a number but never both in UT method.

Let us do a few exercises to reinforce what you have learnt.

Practice Exercise

Find the U of the following multiplication problems. Speed is very important, and you should just see the U without thinking about the T.

Find the U of the following numbers

$$5 \times 3 =$$

$$6 \times 4 =$$

$$9 \times 3 =$$

$$5 \times 2 =$$

$$6 \times 1 =$$

$$7 \times 4 =$$

Here are the answers:

$$5 \times 3 = 5 \text{ (U)}$$

$$6 \times 4 = 4 \text{ (U)}$$

$$9 \times 3 = 7 \text{ (U)}$$

$$5 \times 2 = 0 \text{ (U)}$$

$$6 \times 1 = 6 \text{ (U)}$$

$$7 \times 4 = 8 \text{ (U)}$$

Now let us do the same exercise to find the value of T. Speed is very important and you should just see the T without thinking about the U.

Find the T of the following numbers:

$$5 \times 3 =$$

$$6 \times 4 =$$

$$9 \times 3 =$$

$$5 \times 2 =$$

$$6 \times 1 =$$

$$7 \times 4 =$$

Here are the answers:

$$5 \times 3 = 1 \text{ (T)}$$
$$6 \times 4 = 2 \text{ (T)}$$
$$9 \times 3 = 2 \text{ (T)}$$
$$5 \times 2 = 1 \text{ (T)}$$
$$6 \times 1 = 0 \text{ (T)}$$
$$7 \times 4 = 2 \text{ (T)}$$

Pair Products

Before we look at the UT method, we have to understand what a pair product is.

To find the pair product:

Step 1 - We use the multiplier to multiply each digit of the multiplicand separately.

Step 2 - We then take the U of the product of the left-hand digit of the multiplicand and the T of the product of the right-hand digit of the multiplicand.

Step 3 - We then add the two numbers together to get the pair product.

Let us look at an example to illustrate the method

Example 1
Let us find the pair product of 48 x 6.

$$\begin{array}{r} 48 \\ \underline{\times\ 6} \end{array}$$

Step 1 - We use the multiplier to multiply each digit of the multiplicand separately.

So we multiply 4 by 6 to give 24, and we multiply 8 by 6 to give 48.

$$\begin{array}{r} \mathbf{48} \\ \underline{\mathbf{x\,6}} \\ \mathbf{24 + 48} \end{array}$$

Step 2 - We then take the U of the product of the left-hand digit of the multiplicand and the T of the product of the right-hand digit of the multiplicand.

U of the first digit is 4 and T of the second digit is 4.

$$\begin{array}{r} \mathbf{48} \\ \underline{\mathbf{x\,6}} \\ 2\mathbf{4 + 4}8 \end{array}$$

Step 3 - We then add the two numbers together to get the pair product.

Adding 4 and 4 we get 8 which is the pair product.

$$\begin{array}{r} \mathbf{48} \\ \underline{\mathbf{x\,6}} \\ 2\mathbf{4 + 4}8 \\ \mathbf{8} \end{array}$$

The numbers which are not in boldface are there just for your understanding. When you are visualising in your mind, you should not think about the numbers which are not in boldface. Your focus should just be on the U or the T and adding them to get the pair product.

Example 2
Let us find the pair product of 48 x 5.

48
x 5

Step 1 - We use the multiplier to multiply each digit of the multiplicand separately.

So we multiply 4 by 5 to give 20, and we multiply 8 by 5 to give 40.

48
x 5
20 + 40

Step 2 - We then take the U of the product of the left-hand digit of the multiplicand and the T of the product of the right-hand digit of the multiplicand.

U of the first digit is 0 and T of the second digit is 4.

48
x 5
2**0** + **4**0

Step 3 - We then add the two numbers together to get the pair product.

Adding 0 and 4, we get 4 which is the pair product.

48
x 5
2**0** + **4**0
4

Example 3
Let us find the pair product of 41 x 5.

41
x 5

Step 1 - We use the multiplier to multiply each digit of the multiplicand separately.

So we multiply 4 by 5 to give 20, and we multiply 1 by 5 to give 05.

$$\begin{array}{r} 41 \\ \underline{x\ 5} \\ 20 + 05 \end{array}$$

Step 2 - We then take the U of the product of the left-hand digit of the multiplicand and the T of the product of the right-hand digit of the multiplicand.

U of the first digit is 0 and T of the second digit is 0.

$$\begin{array}{r} 41 \\ \underline{x\ 5} \\ 20 + 05 \end{array}$$

Step 3 - We then add the two numbers together to get the pair product.

Adding 0 and 0, we get 0 which is the pair product.

$$\begin{array}{r} 41 \\ \underline{x\ 5} \\ 20 + 05 \\ 0 \end{array}$$

Example 4
Find the pair product of 28 by 4.

$$\begin{array}{r} 28 \\ \underline{x\ 4} \end{array}$$

Take a second to apply the technique by yourself as fast as you can. Once you have the answer, you can check the steps below to see if you got your answer right.

Step 1 - We use the multiplier to multiply each digit of the multiplicand separately.

So we multiply 2 by 4 to give 08, and we multiply 8 by 4 to give 32.

$$28$$
$$\underline{x\,4}$$
$$08 + 32$$

Step 2 - We then take the U of the product of the left-hand digit of the multiplicand and the T of the product of the right-hand digit of the multiplicand.

U of the first digit is 8 and T of the second digit is 3.

$$28$$
$$\underline{x\,4}$$
$$0\underline{8 + 3}2$$

Step 3 - We then add the two numbers together to get the pair product.

Adding 8 and 3, we get 11 which is the pair product.

$$28$$
$$\underline{x\,4}$$
$$0\underline{8 + 3}2$$
$$\mathbf{11}$$

If you got your answer wrong, don't worry. Just revisit the techniques and examples we covered in this chapter.

Students of the Ofpad Mental Math Course **ofpad.com/mathcourse** can post their questions in the discussion section. If you are not a student of the video course, you can email your questions, suggestions & feedback to **hq@ofpad.com**.

Tips

When applying the UT method:

a) Think of single-digit products as if a 0 is attached in front of it (ex. 3 x 2 = 06).
b) Sometimes when you add U + T, you get a 2 digit number. In such cases, you carry over the one, like how you did in previous methods.
c) Visualize the numbers in your mind and make the pair products.
d) When you feel like you are arriving at the answer without focusing on the steps, the process has become truly automatic.
e) Think about only the U or the T and not the number you are dropping. It will save you time as you do the problem.

Applying UT Method

The rule is simple:

Step 1 – First add the same number of zeros in front of the multiplicand as the number of digits in the multiplier.

Step 2 - Each pair product is a single digit of the answer. In case of two-digit multipliers, the sum of the two pair products is a single digit of the answer.

Let us look at an example.

Example 1

Multiply 4312 by 4

$$4\ 3\ 1\ 2\quad x\ 4$$

Step 1 - Add the same number of zeros as the number of digits in the multiplier.

$$0\ 4\ 3\ 1\ 2\quad x\ 4$$

Step 2 - Each pair product is a single digit of the answer.

Use your middle and forefinger to keep track on whether to calculate U or the T.

$$\begin{array}{l} \text{U T}\\ \underline{0\ 4\ 3\ 1\ 2\quad x\ 4}\\ \_\ \_\ \_\ \_\ \_ \end{array}$$

Pair product of 0 and 4 is 0 + 1 = 1 (0<u>0</u> + <u>1</u>6).

$$\begin{array}{l} \text{U T}\\ \underline{0\ 4\ 3\ 1\ 2\quad x\ 4}\\ 1\ \_\ \_\ \_\ \_ \end{array}$$

Next, the pair product of 4 and 3 is 6 + 1 = 7 (1<u>6</u> + <u>1</u>2)

$$\begin{array}{l} \text{U T}\\ \underline{0\ 4\ 3\ 1\ 2\quad x\ 4}\\ 1\ 7\ \_\ \_\ \_ \end{array}$$

Next, the pair product of 3 and 1 is 2 + 0 = 2 (1<u>2</u> + <u>0</u>4)

$$\begin{array}{l} \text{U T}\\ \underline{0\ 4\ 3\ 1\ 2\quad x\ 4}\\ 1\ 7\ 2\ \_\ \_ \end{array}$$

Next, the pair product of 1 and 2 is 4 + 0 = 4 (0<u>4</u> + <u>0</u>8)

$$\begin{array}{r} \textbf{U T} \\ \underline{\textbf{0 4 3 1 2} \quad \textbf{x 4}} \\ \textbf{1 7 2 4} \_ \end{array}$$

Next, the pair product of 2 and blank is 8 + 0 = 8

$$\begin{array}{r} \textbf{U T} \\ \underline{\textbf{0 4 3 1 2} \quad \textbf{x 4}} \\ \textbf{1 7 2 4 8} \end{array}$$

Now you have the answer 17,248.

The UT Method is not really beneficial for one digit multiplier. You could just use the LR method for one digit multipliers. However, it really speeds up the calculation for two digit multipliers.

Example 2

Let us now try a 2 digit multiplier. Multiply 4312 by 42.

$$\textbf{4 3 1 2} \quad \textbf{x 4 2}$$

Step 1 - Add the same number of zeros as the number of digits in the multiplier.

$$\textbf{0 0 4 3 1 2} \quad \textbf{x 4 2}$$

Step 2 - Sum of two pair products is a single digit of the answer.

Use your middle and forefinger to keep track on whether to calculate U or the T. Place each finger in the middle of two pairs.

$$\begin{array}{l} \textbf{U}_2\,\textbf{T}_2 \\ \quad \textbf{U}_4\,\textbf{T}_4 \\ \underline{\textbf{0 \ 0 \ 4 \ 3 \ 1 \ 2} \quad \textbf{x \ 4 \ 2}} \end{array}$$

Pair product of 0 and 0 is 0 + 0 = 0 (U_2 = 0$\underline{0 + 0}$0).

Pair product of 0 and 4 is 0 + 1 = 1 (U_4 = 0$\underline{0 + 1}$6).

Adding the two pair products 0 + 1 together gives 1.

$$\begin{array}{l} \mathbf{U_2\ T_2} \\ \quad \mathbf{U_4\ T_4} \\ \underline{\mathbf{0\ 0\ 4\ 3\ 1\ 2\quad x\ 4\ 2}} \\ \qquad\quad \mathbf{1}\ \_\ \_\ \_\ \_\ \_ \end{array}$$

Next the pair product of 0 and 4 is 0 + 0 = 0 (U_2 = 0$\underline{0 + 0}$0).

The pair product of 4 and 3 is 6 + 1 = 7 (U_4 = 1$\underline{6 + 12}$).

Adding the two pair products 0 + 7 together gives 7.

$$\begin{array}{l} \mathbf{U_2\ T_2} \\ \quad \mathbf{U_4\ T_4} \\ \underline{\mathbf{0\ 0\ 4\ 3\ 1\ 2\quad x\ 4\ 2}} \\ \qquad \mathbf{1\ 7}\ \_\ \_\ \_\ \_ \end{array}$$

Next the pair product of 4 and 3 is 8 + 0 = 8 (U_2 = 0$\underline{8 + 0}$6).

The pair product of 3 and 1 is 2 + 0 = 2 (U_4 = 1$\underline{2 + 0}$4).

Adding the two pair products 8 + 2 together gives 10.

$$\begin{array}{l} \mathbf{U_2\ T_2} \\ \quad \mathbf{U_4\ T_4} \\ \underline{\mathbf{0\ 0\ 4\ 3\ 1\ 2\quad x\ 4\ 2}} \\ \qquad \mathbf{1\ 7}\ \_\ \_\ \_\ \_ \\ \qquad \mathbf{1\ 0} \end{array}$$

Carrying over the 1, the 7 becomes 8.

$$U_2 T_2$$
$$U_4 T_4$$
$$0\ 0\ 4\ 3\ 1\ 2\quad x\ 4\ 2$$
$$\overline{1\ 8\ 0\ \_\ \_\ \_}$$

Next the pair product of 3 and 1 is 6 + 0 = 6 (U_2 = 0<u>6</u> + 0<u>2</u>).

The pair product of 1 and 2 is 4 + 0 = 4 (U_4 = 0<u>4 + 08</u>).

Adding the two pair products 6 + 4 together gives 10.

$$U_2 T_2$$
$$U_4 T_4$$
$$0\ 0\ 4\ 3\ 1\ 2\quad x\ 4\ 2$$
$$1\ 8\ 0\ \_\ \_\ \_$$
$$+\ 1\ 0$$

Carrying over the 1, so the 0 becomes 1.

$$U_2 T_2$$
$$U_4 T_4$$
$$0\ 0\ 4\ 3\ 1\ 2\quad x\ 4\ 2$$
$$\overline{1\ 8\ 1\ 0\ \_\ \_}$$

Next the pair product of 1 and 2 is 2 + 0 = 2 (U_2 = 0<u>2 + 04</u>).

Pair product of 2 and blank is 8 + 0 = 8 (U_4 = 0<u>8 + 00</u>).

Adding the two pair products together 2 + 8 gives 10.

$$U_2 T_2$$
$$U_4 T_4$$
$$0\ 0\ 4\ 3\ 1\ 2\quad x\ 4\ 2$$
$$1\ 8\ 1\ 0\ \_\ \_$$
$$+1\ 0$$

Carrying over the 1, the 0 becomes 1.

$$U_2\ T_2$$
$$U_4\ T_4$$
$$\underline{0\ 0\ 4\ 3\ 1\ 2\quad x\ 4\ 2}$$
$$1\ 8\ 1\ 1\ 0\ \_$$

Next the pair product of 2 and blank is 4 + 0 = 4 ($U_2 = 0\underline{4}\ +$ $\underline{0}0$) and that becomes the last digit of the answer.

$$U_2\ T_2$$
$$U_4\ T_4$$
$$\underline{0\ 0\ 4\ 3\ 1\ 2\qquad x\ 4\ 2}$$
$$1\ 8\ 1\ 1\ 0\ 4$$

Now you have the answer 181,104

Example 3
Let us try another example. Multiply 9238 by 84.

$$9\ 2\ 3\ 8\quad x\ 8\ 4$$

Step 1 - Add the same number of zeros as the number of digits in the multiplier.

$$\underline{0\ 0\ 9\ 2\ 3\ 8\quad x\ 8\ 4}$$

$$-\ -\ -\ -\ -\ -$$

Step 2 - Sum of two pair products is a single digit of the answer.

Use your middle and forefinger to keep track on whether to calculate U or the T. Place each finger in the middle of two pairs.

$$U_4\ T_4$$
$$U_8\ T_8$$
$$\underline{0\ 0\ 9\ 2\ 3\ 8\quad x\ 8\ 4}$$

Pair product of 0 and 0 is 0 + 0 = 0 (U_4 = 0<u>0</u> + <u>0</u>0).

Pair product of 0 and 9 is 0 + 7 = 7 (U_8 = 0<u>0</u> + <u>7</u>2).

Adding the two together gives 7.

$$U_4\ T_4$$
$$U_8\ T_8$$
$$\underline{0\ \ 0\ \ 9\ \ 2\ \ 3\ \ 8}\quad \text{x}\,8\,4$$
$$7\ \_\ \_\ \_\ \_$$

Next the pair product of 0 and 9 is 0 + 3 = 3 (U_4 = 0<u>0</u> + <u>3</u>6).

Pair product of 9 and 2 is 2 + 1 = 3 (U_8 = 7<u>2</u> + <u>1</u>6).

Adding the two together gives 6.

$$U_4\ T_4$$
$$U_8\ T_8$$
$$\underline{0\ \ 0\ \ 9\ \ 2\ \ 3\ \ 8}\quad \text{x}\,8\,4$$
$$7\ 6\ \_\ \_\ \_\ \_$$

Next the pair product of 9 and 2 is 6 + 0 = 6 (U_4 = 3<u>6</u> + <u>0</u>8).

Pair product of 2 and 3 is 6 + 2 = 8 (U_8 = 1<u>6</u> + <u>2</u>4).

Adding the two together gives 14.

$$U_4\ T_4$$
$$U_8\ T_8$$
$$\underline{0\ \ 0\ \ 9\ \ 2\ \ 3\ \ 8}\quad \text{x}\,8\,4$$
$$7\ 6\ \_\ \_\ \_\ \_$$
$$+\,1\,4$$

Carrying over the 1, the 6 becomes 7.

$$\underline{0\ \ 0\ \ 9\ \ 2\ \ 3\ \ 8}\quad \text{x}\,8\,4$$
$$7\ 7\ 4\ \_\ \_\ \_$$

Next the pair product of 2 and 3 is 8 + 1 = 9 (U_4 = 0$\underline{8}$ + $\underline{1}$2).

Pair product of 3 and 8 is 4 + 6 = 10 (U_8 = 2$\underline{4}$ + $\underline{6}$4).

Adding the two together gives 19.

$$\begin{array}{r} \mathbf{U_4\ T_4} \\ \mathbf{U_8\ T_8} \\ \underline{\mathbf{0\ 0\ 9\ 2\ 3\ 8} \quad \mathbf{x\ 8\ 4}} \\ \mathbf{7\ 7\ 4}\ \_\ \_\ \_ \\ \mathbf{+\ 1\ 9} \end{array}$$

Carrying over the 1, the 4 becomes 5.

$$\begin{array}{r} \underline{\mathbf{0\ 0\ 9\ 2\ 3\ 8} \quad \mathbf{x\ 8\ 4}} \\ \mathbf{7\ 7\ 5\ 9}\ \_\ \_ \end{array}$$

Next the pair product of 3 and 8 is 2 + 3 = 5 (U_4 = 1$\underline{2}$ + $\underline{3}$2).

Pair product of 8 and blank is 4 + 0 = 4 (U_8 = 6$\underline{4}$ + $\underline{0}$0).

Adding the two together gives 9.

$$\begin{array}{r} \mathbf{U_4\ T_4} \\ \mathbf{U_8\ T_8} \\ \underline{\mathbf{0\ 0\ 9\ 2\ 3\ 8} \quad \mathbf{x\ 8\ 4}} \\ \mathbf{7\ 7\ 5\ 9\ 9}\ \_ \end{array}$$

Next the pair product of 8 and blank is 2 + 0 = 2 (U_4 = 3$\underline{2}$ + $\underline{0}$0) and that becomes the last digit of the answer.

$$\begin{array}{r} \mathbf{U_4\ T_4} \\ \mathbf{U_8\ T_8} \\ \underline{\mathbf{0\ 0\ 9\ 2\ 3\ 8} \quad \mathbf{x\ 8\ 4}} \\ \mathbf{7\ 7\ 5\ 9\ 9\ 2} \end{array}$$

Now you have the answer 775,992.

Example 4
Multiply 5743 by 63.

$$\mathbf{5\,7\,4\,3 \quad x\,6\,3}$$

Take a second to apply the technique by yourself as fast as you can. Once you have the answer, you can check the steps below to see if you got your answer right.

Step 1 - Add the same number of zeros as the number of digits in the multiplier.

$$\underline{\mathbf{0\ 0\ 5\ 7\ 4\ 3 \quad x\,6\,3}}$$
$$\text{– – – – – –}$$

Step 2 - Sum of two pair products is a single digit of the answer.

$$\mathbf{U_3\,T_3}$$
$$\mathbf{U_6\,T_6}$$
$$\underline{\mathbf{0\ 0\ 5\ 7\ 4\ 3 \quad x\,6\,3}}$$
$$\text{– – – – – –}$$

Use your middle and forefinger to keep track on whether to calculate U or the T. Place each finger in the middle of two pairs.

Pair product of 0 and 0 is $0 + 0 = 0$ ($U_3 = 0\underline{0} + \underline{0}0$).

Pair product of 0 and 5 is $0 + 3 = 3$ ($U_6 = 0\underline{0} + \underline{3}0$).

Adding the two together gives 3.

$$\mathbf{U_3\,T_3}$$
$$\mathbf{U_6\,T_6}$$
$$\underline{\mathbf{0\ 0\ 5\ 7\ 4\ 3 \quad x\,6\,3}}$$
$$\mathbf{3}\ \text{– – – – –}$$

Next the pair product of 0 and 5 is 0 + 1 = 1 ($U_3 = \underline{00} + \underline{1}5$).

Pair product of 5 and 7 is 0 + 4 = 4 ($U_6 = \underline{3}0 + \underline{4}2$).

Adding the two together gives 5.

$$\mathbf{U_3\ T_3}$$
$$\mathbf{U_6\ T_6}$$
$$\mathbf{0\ \ 0\ \ 5\ \ 7\ \ 4\ \ 3\ \ \ \ x\ 6\ 3}$$
$$\mathbf{3\ 5\ \_\ \_\ \_\ \_}$$

Next the pair product of 5 and 7 is 5 + 2 = 7 ($U_3 = 1\underline{5} + \underline{2}1$).

Pair product of 7 and 4 is 2 + 2 = 4 ($U_6 = 4\underline{2} + \underline{2}4$).

Adding the two together gives 11.

$$\mathbf{U_3\ T_3}$$
$$\mathbf{U_6\ T_6}$$
$$\mathbf{0\ \ 0\ \ 5\ \ 7\ \ 4\ \ 3\ \ \ \ x\ 6\ 3}$$
$$\mathbf{3\ 5\ \_\ \_\ \_\ \_}$$
$$\mathbf{+\ 1\ 1}$$

Carrying over the 1, the 5 becomes 6.

$$\mathbf{0\ \ 0\ \ 5\ \ 7\ \ 4\ \ 3\ \ \ \ x\ 6\ 3}$$
$$\mathbf{3\ 6\ 1\ \_\ \_\ \_}$$

Next the pair product of 7 and 4 is 1 + 1 = 2 ($U_3 = 2\underline{1} + \underline{1}2$).

Pair product of 4 and 3 is 4 + 1 = 5 ($U_6 = 2\underline{4} + \underline{1}8$).

Adding the two together gives 7.

$$\mathbf{U_3\ T_3}$$
$$\mathbf{U_6\ T_6}$$
$$\mathbf{0\ \ 0\ \ 5\ \ 7\ \ 4\ \ 3\ \ \ \ x\ 6\ 3}$$
$$\mathbf{3\ 6\ 1\ 7\ \_\ \_}$$

Next the pair product of 4 and 3 is 2 + 0 = 2 (U_3 = 1<u>2</u> + <u>0</u>9).

Pair product of 3 and blank is 8 + 0 = 8 (U_6 = 1<u>8</u> + <u>0</u>0).

Adding the two together gives 10.

$$\textbf{U}_3\ \textbf{T}_3$$
$$\textbf{U}_6\ \textbf{T}_6$$
$$\underline{\textbf{0 0 5 7 4 3}\quad \textbf{x 6 3}}$$
$$\textbf{3 6 1 7} \_ \_$$
$$\textbf{+ 1 0}$$

Carrying over the 1, the 7 becomes 8.
$$\underline{\textbf{0 0 5 7 4 3}\quad \textbf{x 6 3}}$$
$$\textbf{3 6 1 8 0} \_$$

Next the pair product of 3 and blank is 9 + 0 = 9 (U_3 = <u>0</u>9 + <u>0</u>0) and that becomes the last digit of the answer.

$$\textbf{U}_3\ \textbf{T}_3$$
$$\textbf{U}_6\ \textbf{T}_6$$
$$\underline{\textbf{0 0 5 7 4 3}\quad \textbf{x 6 3}}$$
$$\textbf{3 6 1 8 0 9}$$

Now you have the answer 361,809.

If you got your answer wrong, don't worry. Just revisit the techniques and examples we covered in this chapter.

Students of the Ofpad Mental Math Course **ofpad.com/mathcourse** can post their questions in the discussion section. If you are not a student of the video course, you can email your questions, suggestions & feedback to **hq@ofpad.com**.

Read this chapter again if necessary.

Then go to the practice section and complete the exercises.

You might have understood the technique, but only practice will make using these techniques effortless and easy.

If you have enjoyed the book so far, do leave a review on Amazon by visiting **ofpad.com/mathbook**, so that others might also benefit from reading this book.

Exercises
Download the rich PDFs for these exercises from
ofpad.com/mathexercises.

01) 2951 x 7	07) 7469 x 96	13) 6671 x 386
02) 1315 x 7	08) 7770 x 73	14) 6402 x 290
03) 5127 x 6	09) 8626 x 69	15) 2185 x 939
04) 2934 x 9	10) 4025 x 70	16) 1919 x 803
05) 6999 x 8	11) 7731 x 63	17) 9372 x 646
06) 7326 x 7	12) 3196 x 63	18) 3374 x 999

Answers
01) 20,657	07) 717,024	13) 257,006
02) 9,205	08) 567,210	14) 1,856,580
03) 30,762	09) 586,568	15) 2,051,715
04) 26,406	10) 281,750	16) 1,540,957
05) 55,992	11) 487,053	17) 6,054,312
06) 51,282	12) 201,348	18) 3,370,626

Chapter 14 - FP Division

In this chapter, we will look at how to do fast division using the flagpole method. This method might require you to write down the numbers, but with practice, you will be able to solve it without doing that.

Flag & Pole

Before we apply the flagpole method to division, it is important to understand which digit is a flag and which digit is a pole.

a) The first or the first set of numbers is the pole.
b) The second or the second set of numbers is the flag.

Let's look at a few numbers.

In 25, the pole is 2, and the flag is 5. It can be written like this 2^5.

In 83, the pole is 8, and the flag is 3. It can be written like this 8^3.

In 123, the pole is 12, and the flag is 3. It can be written like this 12^3.

Depending on how you group the number, in 123, you can also make 1 the pole and 23 as the flag. So it can also be written like this 1^{23}.

Applying Flag Pole Method

To apply the flagpole method to division.

Step 1 - Divide the first digit of the dividend by the pole to find the first digit of the answer.

Step 2 - Attach the remainder from the previous step to the next digit of the dividend.

Step 3 - Multiply the last calculated digit of the answer with the flag and subtract it from the number arrived at after attaching the remainder.

Step 3b - If the number after subtraction is negative, reduce the last calculated digit of your answer by 1 and add the pole to the remainder before attaching it to the next digit of the dividend, and then repeat the previous step.

Step 4 - Divide the number from the previous step, by the pole to get the next digit of the answer.

Step 5 - If you cross the decimal line in the dividend add a decimal point.

Repeat steps 3 and 4 until all the digits of the answer is calculated.

Example 1
Let us now look at a few examples.

Divide 5578 by 25.

$$25\underline{|\ 5\ \ 5\ \ 7\ |\ 8}$$

$$-\ -\ -\ -$$

So here 2 is the pole, and 5 is the flag.

$$2^5\underline{|\ 5\ \ \ 5\ \ 7\ |\ 8}$$

$$-\ -\ -\ -$$

Step 1 - Divide the first digit of the dividend by the pole to find the first digit of the answer.

Divide 5 by 2 to get 2 with remainder 1

$$2^5 | 5 \quad 5 \quad 7 | 8$$
$$2 \_ \_ \_$$

Step 2 - Attach the remainder from the previous step to the next digit of the dividend.

So attach 1 to the next digit.

$$2^5 | 5 \quad 15 \quad 7 | 8$$
$$2 \_ \_ \_$$

Step 3 - Multiply the last calculated digit of the answer with the flag and subtract it from the number arrived at after attaching the remainder.

So multiply 2 with the flag 5 to get 10. Subtract it from 15 to get 5.

Step 4 - Divide the number from the previous step, by the pole to get the next digit of the answer.

We divide 5 by the pole 2 to get the next digit of the answer 2, and the remainder 1 is attached to the next digit.

$$2^5 | 5 \quad 5 \quad 17 | 8$$
$$2 \, 2 \_ \_$$

Repeat steps 3 and 4 until all the digits of the answer is calculated.

Step 3 - Multiply the last calculated digit of the answer with the flag and subtract it from the number arrived at after attaching the remainder.

So multiply 2 with the flag 5 to get 10. Subtract it from 17 to get 7.

Step 4 - Divide the number from the previous step, by the pole to get the next digit of the answer.

We divide 7 by the pole 2 to get the next digit of the answer 3, and the remainder 1 is attached to the next digit.

$$2^5 | \; 5 \; 5 \; 17 \; |8$$
$$2 \; 2 \; 3 \; \_$$

Step 5 - If you cross the decimal line in the dividend add a decimal point.

Since we are crossing the decimal line, let us add a decimal point before carrying on.

$$2^5 | \; 5 \; 5 \; 7 \; |8$$
$$2 \; 2 \; 3 \; . \; \_$$

Repeat steps 3 and 4 until all the digits of the answer is calculated.

Step 3 - Multiply the last calculated digit of the answer with the flag and subtract it from the number arrived at after attaching the remainder.

So multiply 3 with the flag 5 to get 15. Subtract it from 18 to get 3.

Step 4 - Divide the number from the previous step, by the pole to get the next digit of the answer.

We divide 3 by the pole 2 to get the next digit of the answer 1, and the remainder 1 is attached to the next digit.

$$2^5 | \ 5 \ \ 5 \ \ 7 \ |8 \ 1$$
$$2 \ 2 \ 3 \ . \ 1$$

We add an extra zero since there are no more digits, and we have crossed the decimal line.

$$2^5 | \ 5 \ \ 5 \ \ 7 \ |8 \ 10$$
$$2 \ 2 \ 3 \ . \ 1$$

Repeat steps 3 and 4 until all the digits of the answer is calculated.

Step 3 - Multiply the last calculated digit of the answer with the flag and subtract it from the number arrived at after attaching the remainder.

So multiply 1 with the flag 5 to get 5. Subtract it from 10 to get 5.

Step 4 - Divide the number from the previous step, by the pole to get the next digit of the answer.

We divide 5 by the pole 2 to get 2. If you want to calculate additional decimal points, then you can carry on this way by attaching the remainder 1 to the next digit.

$$2^5 | \ 5 \ \ 5 \ \ 7 \ |8 \ 10$$
$$2 \ 2 \ 3 \ . \ 1 \ 2$$

Example 2

Let us now look at another example.

Divide 601,324 by 76.

$$76 \underline{\lfloor\, 6\ 0\ 1\ 3\ 2\ \lfloor\, 4}$$

So here 7 is the pole, and 6 is the flag.

$$7^6 \underline{\lfloor\, 6\ 0\ 1\ 3\ 2\ \lfloor\, 4}$$

Step 1 - Divide the first digit of the dividend by the pole to find the first digit of the answer.

Divide 60 by 7 to get 8 with remainder 4.

$$7^6 \underline{\lfloor\, 6\ 0\ 1\ 3\ 2\ \lfloor\, 4}$$
$$8\ \_\ \_\ \_\ \_\ \_$$

Step 2 - Attach the remainder from the previous step to the next digit of the dividend.

So attach 4 to the next digit.

$$7^6 \underline{\lfloor\, 6\ 0\ {}_4 1\ 3\ 2\ \lfloor\, 4}$$
$$8\ \_\ \_\ \_\ \_\ \_$$

Step 3 - Multiply the last calculated digit of the answer with the flag and subtract it from the number arrived at after attaching the remainder.

So multiply 8 with the flag 6 to get 48. Subtract it from 41 to get -7.

Step 3b - If the number after subtraction is negative, reduce the last calculated digit of your answer by 1 and add the pole to the remainder before attaching it to the next digit of the dividend, and then repeat the previous step.

So we reduce 8 by 1 to get 7.

$$7^6 | \ 6 \ 0 \ _{41}1 \ 3 \ 2 \ | \ 4$$
$$7 \ \_ \ \_ \ \_ \ \_ \ \_$$

We add the pole 7 to the previous remainder 4 to get 11 and attach it to the next digit.

$$7^6 | \ 6 \ 0 \ _{111}1 \ 3 \ 2 \ | \ 4$$
$$7 \ \_ \ \_ \ \_ \ \_ \ \_$$

Step 3 - Multiply the last calculated digit of the answer with the flag and subtract it from the number arrived at after attaching the remainder.

So we multiply 7 with the flag 6 to get 42. Subtract it from 111 to get 69.

Step 4 - Divide the number from the previous step, by the pole to get the next digit of the answer.

We divide 69 by the pole 7 to get the next digit of the answer 9, and the remainder 6 is attached to the next digit.

$$7^6 | \ 6 \ 0 \ 1 \ _{63}3 \ 2 \ | \ 4$$
$$7 \ 9 \ \_ \ \_ \ \_ \ \_$$

Repeat steps 3 and 4 until all the digits of the answer is calculated.

Step 3 - Multiply the last calculated digit of the answer with the flag and subtract it from the number arrived at after attaching the remainder.

So multiply 9 with the flag 6 to get 54. Subtract it from 63 to get 9.

Step 4 - Divide the number from the previous step, by the pole to get the next digit of the answer.

We divide 9 by the pole 7 to get the next digit of the answer 1, and the remainder 2 is attached to the next digit.

$$7^6| \; 6 \; 0 \; 1 \; 63 \; 22 \; | \; 4$$
$$\textbf{7 \; 9 \; 1} \_ \; \_ \; \_$$

Repeat steps 3 and 4 until all the digits of the answer is calculated.

Step 3 - Multiply the last calculated digit of the answer with the flag and subtract it from the number arrived at after attaching the remainder.

So multiply 1 with the flag 6 to get 6. Subtract it from 22 to get 16.

Step 4 - Divide the number from the previous step, by the pole to get the next digit of the answer.

We divide 16 by the pole 7 to get the next digit of the answer 2, and the remainder 2 is attached to the next digit.

$$7^6| \; 6 \; 0 \; 1 \; 3 \; 2 \; | 24$$
$$\textbf{7 \; 9 \; 1 \; 2} \_ \; \_$$

Step 5 - If you cross the decimal line in the dividend add a decimal point.

Since we are crossing the decimal line, let us add a decimal point before carrying on.

$$7^6 | \; 6 \; 0 \; 1 \; 3 \; 2 \; | 24$$
$$7 \; 9 \; 1 \; 2 . \_ \; \_$$

Repeat steps 3 and 4 until all the digits of the answer is calculated.

Step 3 - Multiply the last calculated digit of the answer with the flag and subtract it from the number arrived at after attaching the remainder.

So multiply 2 with the flag 6 to get 12. Subtract it from 24 to get 12.

Step 4 - Divide the number from the previous step, by the pole to get the next digit of the answer.

We divide 12 by the pole 7 to get the next digit of the answer 1, and the remainder 5 is attached to the next digit.

$$7^6 | \; 6 \; 0 \; 1 \; 3 \; 2 \; | 24 \; 5$$
$$7 \; 9 \; 1 \; 2 . 1 \; \_$$

We add an extra zero since there are no more digits.

$$7^6 | \; 6 \; 0 \; 1 \; 3 \; 2 \; | 4 \; 50$$
$$7 \; 9 \; 1 \; 2 . 1 \; \_$$

Repeat steps 3 and 4 until all the digits of the answer is calculated.

Step 3 - Multiply the last calculated digit of the answer with the flag and subtract it from the number arrived at after attaching the remainder.

So multiply 1 with the flag 6 to get 6. Subtract it from 50 to get 44.

Step 4 - Divide the number from the previous step, by the pole to get the next digit of the answer.

We divide 44 by the pole 7 to get the next digit of the answer 6.

$$7^6| \ 6 \ 0 \ 1 \ 3 \ 2 \ |4 \ 50$$
$$7 \ 9 \ 1 \ 2.16$$

If you want to calculate additional decimal points, then you can carry on this way by attaching the remainder to the next digit.

Example 3

Let us now look at another example.

Divide 2829 by 123.

$$123| \ 2 \ 8 \ 2 \ | \ 9$$
$$\_ \ \_ \ \_ \ \_ \ \_$$

So here 12 is the pole, and 3 is the flag.

Step 1 - Divide the first digit of the dividend by the pole to find the first digit of the answer

Divide 28 by 12 to get 2 with remainder 4

$$12^3| \ 2 \ 8 \ 2 \ | \ 9$$
$$2 \_ \ \_ \ \_$$

Step 2 - Attach the remainder from the previous step to the next digit of the dividend.

$$12^3| \ 2 \ 8 \ 42 \ | \ 9$$
$$2 \_ \ \_ \ \_$$

Step 3 - Multiply the last calculated digit of the answer with the flag and subtract it from the number arrived at after attaching the remainder.

So multiply 2 with the flag 3 to get 6. Subtract it from 42 to get 36.

Step 4 - Divide the number from the previous step, by the pole to get the next digit of the answer.

We divide 36 by the pole 12 to get the next digit of the answer 3, and the remainder 0 is attached to the next digit.

$$12^3 | \ 2 \ 8 \ 2 \ | \ 09$$
$$2 \ 3 \ \_ \ \_$$

Step 5 - If you cross the decimal line in the dividend add a decimal point.

Since we are crossing the decimal line, let us add the decimal point.

$$12^3 | \ 2 \ 8 \ 2 \ | \ 09$$
$$2 \ 3 \ . \ \_$$

Repeat steps 3 and 4 until all the digits of the answer is calculated.

Step 3 - Multiply the last calculated digit of the answer with the flag and subtract it from the number arrived at after attaching the remainder.

So multiply 3 with the flag 3 to get 9. Subtract it from 9 to get 0.

Step 4 - Divide the number from the previous step, by the pole to get the next digit of the answer.

We divide 0 by the pole 12 to get the next digit of the answer
0

$$12^3 | 2\ 8\ 2 | 09$$
$$2\ 3\ .\ 0$$

Example 4
Let us now look at another example.

Divide 3517 by 127.

$$127 | 3\ 5\ 1\ | 7$$
$$-\ -\ -\ -\ -$$

Take a second to apply the technique by yourself as fast as you can. Once you have the answer, you can check the steps below to see if you got your answer right.

So here 12 is the pole, and 7 is the flag.

$$12^7 | 3\ 5\ 1\ | 7$$
$$-\ -\ -\ -$$

Step 1 - Divide the first digit of the dividend by the pole to find the first digit of the answer.

Divide 35 by 12 to get 2 with remainder 11.

$$12^7 | 3\ 5\ 1\ | 7$$
$$2\ \_\ \_\ \_$$

Step 2 - Attach the remainder from the previous step to the next digit of the dividend.

So attach 11 to the next digit.

$$12^{7}| \ 3 \ _{11}5 \ 1 \ | \ 7$$
$$2 \_ \_ \_$$

Step 3 - Multiply the last calculated digit of the answer with the flag and subtract it from the number arrived at after attaching the remainder.

So multiply 2 with the flag 7 to get 14. Subtract it from 111 to get 97.

Step 4 - Divide the number from the previous step, by the pole to get the next digit of the answer.

We divide 97 by the pole 12 to get the next digit of the answer 8, and the remainder 1 is attached to the next digit.

$$12^{7}| \ 3 \ 5 \ _{1}11 \ | \ 17$$
$$2 \ 8 \_ \_$$

Step 5 - If you cross the decimal line in the dividend add a decimal point.

We are crossing the decimal line so let us add a decimal point.

$$12^{7}| \ 3 \ 5 \ 1 \ | \ 17$$
$$2 \ 8 \ . \_ \_$$

Repeat steps 3 and 4 until all the digits of the answer is calculated.

Step 3 - Multiply the last calculated digit of the answer with the flag and subtract it from the number arrived at after attaching the remainder.

So multiply 8 with the flag 7 to get 56. Subtract it from 17 to get -39.

Step 3b - If the number after subtraction is negative, reduce the last calculated digit of your answer by 1 and add the pole to the remainder before attaching it to the next digit of the dividend, and then repeat the previous step.

So we reduce 8 by 1 to get 7.

We add the pole 12 to the remainder 1 to get 13 and attach it to the next digit.

$$12^7| \, 3 \; 5 \; 1 \; | \; 13\underline{7}$$
$$2 \; 7 \; . \; \_ \; \_$$

Step 3 - Multiply the last calculated digit of the answer with the flag and subtract it from the number arrived at after attaching the remainder.

So we multiply 7 with the flag 7 to get 49. Subtract it from 137 to get 88.

Step 4 - Divide the number from the previous step, by the pole to get the next digit of the answer.

We divide 88 by the pole 12 to get the next digit of the answer 7, and the remainder 4 is attached to the next digit.

$$12^7| \, 3 \; 5 \; 1 \; | \; 13\underline{7} \, 4$$
$$2 \; 7 \; . \; 7 \; \_$$

Attach an extra 0 since we crossed the decimal, and there are no more digits left.

$$12^7| \, 3 \; 5 \; 1 \; | \; 7 \, 40$$
$$2 \; 7 \; . \; 7 \; \_$$

Repeat steps 3 and 4 until all the digits of the answer is calculated.

Step 3 - Multiply the last calculated digit of the answer with the flag and subtract it from the number arrived at after attaching the remainder.

So multiply 7 with the flag 7 to get 49. Subtract it from 40 to get -9.

Step 3b - If the number after subtraction is negative, reduce the last calculated digit of your answer by 1 and add the pole to the remainder before attaching it to the next digit of the dividend, and then repeat the previous step.

So we reduce 7 by 1 to get 6.

We add the pole 12 to the remainder 4 to get 16 and attach it to the next digit.

$$12^7 | \ 3\ 5\ 1\ |\ 7\ 160$$
$$2\ 7\ .\ 6\ \_$$

Step 3 - Multiply the last calculated digit of the answer with the flag and subtract it from the number arrived at after attaching the remainder.

Then we multiply 6 with the flag 7 to get 42. Subtract it from 160 to get 118.

Step 4 - Divide the number from the previous step, by the pole to get the next digit of the answer.

We divide 118 by the pole 12 to get the next digit of the answer 9, and the remainder 10 is attached to the next digit.